T0140856

Xiaowei Zhang

Supporting the Understanding of Rare Disease Diagnostics with Questionnaire-Based Data Analysis and Computer-Aided Classifier Fusion

Logos Verlag Berlin

λογος

Bibliografische Information der Deutschen Nationalbibliothek

Die Deutsche Nationalbibliothek verzeichnet diese Publikation in der Deutschen Nationalbibliografie; detaillierte bibliografische Daten sind im Internet über http://dnb.d-nb.de abrufbar.

ISBN 978-3-8325-5668-6

Logos Verlag Berlin GmbH
Georg-Knorr-Str. 4, Geb. 10,
D-12681 Berlin
Germany

Tel.: +49 (0)30 / 42 85 10 90
Fax: +49 (0)30 / 42 85 10 92
http://www.logos-verlag.de

Abstract

Orphan diseases are characterized by low incidence and complex pathology. The conventional examination methods are arduous but cannot guarantee the correctness of a diagnosis. The limited acquaintance with and knowledge about rare disorders, coupled with insufficient epidemiological evidence and clinical experiences, makes it challenging for general practitioners to identify rare diseases reliably. Medication mistakes based on an inaccurate diagnosis may cause adverse effects on the patient's health. With advancements in artificial intelligence, decision support systems are gradually gaining prominence in modern medical diagnosis. An innovative medical service model centered on patient symptoms provides technological advantages in improving the future direction of orphan disease diagnostic and treatment processes.

This research aims to amass a rare disease repository by collecting diagnosed patients' information and complementary therapeutic schemes, thereby providing elaborate references for disease diagnostics and reducing gaps in the interpretation of rare disease symptoms. A prospective study was created in cooperation with Medizinische Hochschule Hannover. The interdisciplinary research focuses on developing a computer-aided diagnostic support system based on statistical analysis and machine learning algorithms to optimize rare disease diagnosis processes. An ensemble of classification methods was conceptualized to discover the potential rules inherent in the patient questionnaire data, thus incorporating disease profile aggregation, pattern recognition, and information comparison. Moreover, an interactive data visualization platform was successively established, promoting intuitive understanding through diagnosis graphics that depict orphan diseases' specific characteristics. Data visualization strengthens the transmission efficiency of important information, motivates disease researchers to explore data, and provides in-depth insights into problematic connected symptoms. The practically realized concept reveals evidence to sensitize doctors to correct decision-making based on an internal evaluation for system-aided diagnosis. A self-adapting questioning mechanism results in personalized questionnaires that automatically analyze suspicious disease trends and accumulate experience.

An iterative validation process reveals that the classifier fusion method presented notable performance improvements compared to any single supervised classification model. The patient-oriented inquiry algorithm can efficiently reduce unnecessary questions while having a reliable diagnosis. By effectively combining theory with practice, the statistical learning method combined with the visualization module can systematically present disease-related symptoms patterns, thereby guiding medical practitioners in increasing their understanding of orphan disease characteristics, and providing new means for rare disorder diagnostics based on the system's learning results from previously diagnosed patients. The supplementary diagnosis prediction mechanism introduced in this doctoral dissertation is not confined to questionnaires on orphan diseases. Indeed, the methodological procedure can be applied to any survey with closed-ended questions for researchers interested in analyzing the predispositions and significant influence factors of different groups.

Zusammenfassung

Seltene Krankheiten haben eine geringe Inzidenz und eine komplexe Pathologie. Die herkömmlichen Untersuchungsmethoden sind mühsam, können jedoch die Richtigkeit einer Diagnose nicht immer garantieren. Begrenztes Wissen und Kenntnisse über seltene Erkrankungen in Verbindung mit unzureichenden epidemiologischen Nachweisen und klinischen Erfahrungen machen es für Allgemeinmediziner schwierig, seltene Krankheiten zuverlässig zu identifizieren. Falsche oder suboptimale Medikationen, die auf einer ungenauen Diagnose zurückzuführen sind, können sich nachteilig auf die Gesundheit des Patienten auswirken. Im Zeitalter der computergestützten künstlichen Intelligenz werden Entscheidungssysteme zunehmend als Assistenz in der modernen medizinischen Diagnose verwendet. Das in dieser Arbeit entwickelte medizinische Servicemodell, das sich auf Patientensymptome konzentriert, bietet die technologischen innovativen Vorteile, um die zukünftige Ausrichtung für Diagnoseprozesse für seltene Erkrankungen und die Behandlung von Patienten zu verbessern.

Diese Forschung zielt darauf ab, eine zentrale Wissensbasis für seltene Krankheiten aufzubauen, indem Informationen diagnostizierter Patienten und ergänzende therapeutische Schemata zentral gesammelt werden, wodurch wertvolle Referenzen für die Diagnostik von Krankheiten bereitgestellt und Lücken bei der Interpretation seltener Krankheitssymptome geschlossen werden. In Kooperation mit der Medizinischen Hochschule Hannover wurde eine prospektive Studie durchgeführt. Die interdisziplinäre Forschung konzentriert sich auf die Entwicklung eines computergestützten diagnostischen Unterstützungssystems, das auf statistischen Analysen und Algorithmen für maschinelles Lernen basiert, um Prozesse zur Diagnose seltener Krankheiten zu optimieren. Ein Ensemble von Klassifizierungsmethoden wurde konzipiert, um die potenziellen Symptommuster zu ermitteln, die aus Daten eines Patientenfragebogens der MHH hervorgehen, und so die systematische Aggregation von Krankheitsprofilen, die Mustererkennung und den Informationsvergleich berücksichtigen. In diesem Zusammenhang wurde nacheinander eine interaktive Datenvisualisierungsplattform entwickelt, die ein intuitives Verständnis durch Diagnosegrafiken fördert, die die spezifischen Merkmale von seltenen Erkrankungen darstellen. Die Datenvisualisierung stärkt die Übertragungseffizienz wichtiger Informationen, motiviert Krankheitsforscher zur Erforschung von Daten und bietet detaillierte Einblicke in Problem-verbundene Symptome. Das praktisch realisierte Konzept ermöglicht belegbare Diagnosen, um Ärzte für eine korrekte Entscheidungsfindung auf der Grundlage einer internen Bewertung für eine systemgestützte Diagnose zu sensibilisieren. Ein dazu entwickelter Fragenalgorithmus führt zu personalisierten Fragebögen basierend auf verdächtigen Krankheitstrends, welche automatisch analysiert werden, um nicht relevante Symptomfragen für den Patienten auszugrenzen.

Ein iterativer Validierungsprozess zeigt, dass die Classifier-Fusion-Methode im Vergleich zu einem einzelnen überwachten Klassifizierungsmodell bemerkenswerte Leistungsverbesserungen aufweist. Der patientenorientierte Fragenstellungsalgorithmus kann unnötige Fragen effektiv reduzieren und gleichzeitig eine zuverlässige Diagnose liefern. Durch die effektive Kombination von Theorie und Praxis kann die statistische Lernmethode in Kombination mit dem Visualisierungsmodul krankheitsbezogene Symptommuster systematisch darstellen, wodurch Ärzte bei der Verständnisverbesserung der Merkmale seltener Krankheiten angeleitet werden und neue Mittel für die Diagnose seltener Erkrankungen auf der Grundlage des Systems bereitgestellt werden Lernergebnisse von zuvor diagnostizierten Patienten. Der in dieser Dissertation vorgestellte ergänzende Diagnosevorhersagemechanismus ist nicht auf Fragebögen zu seltenen Krankheiten beschränkt. Tatsächlich kann das methodische Vorgehen auf jede Umfrage mit geschlossenen Fragen für Forscher angewendet werden, die daran interessiert sind, die Prädispositionen und signifikanten Einflussfaktoren verschiedener Gruppen zu analysieren.

Contents

List of Figures

List of Tables

Abbreviations

AI	Artificial Intelligence
ALS	Amyotrophic Lateral Sclerosis
ANNs	Artificial Neural Networks
ANOVA	Analysis of Variance
AUC	Area Under the ROC Curve
BC	Breast Cancer
CCCV	Calibrated-Classifier Cross-Validation
CD	Cronic Diseases
CPGs	Clinical Practice Guidelines
CPP	Calibrated Prediction Probability
DTs	Decision Trees
EBM	Evidence-Based Medicine
FN	False Negative
FP	False Positive
FPR	False Positive Rate
GANs	Generative Adversarial Networks
GDA	Generalized Discriminant Analysis
GMM	Gaussian Mixture Models
HMM	Hidden Markov Models
KDD	Knowledge Discovery in Databases
KMS	Knowledge Management System
KNN	K-Nearest Neighbors
LDA	Linear Discriminant Analysis
LOOCV	Leave-one-out Cross-Validation
Log-Loss	Logarithmic Loss or Cross-Entropy Loss
LR	Logistic Regression
LS-SVM	Lease Square Support Vector Machine
MAE	Mean Absolute Error
MDs	Medical Doctors
MHH	Medizinische Hochschule Hannover
MI	Mutual Information
MIC	Maximal Information Coefficient
ML	Machine Learning
MSE	Mean Squared Error

NB	Naïve Bayes
NPV	Negative Predictive Value
NRDs	Non-Rare Diseases
PAVA	Pool Adjacent Violators Algorithm
PCA	Principal Components Analysis
PID	Primary Immunodeficiency Disorders
PPV	Positive Predictive Value
PR	Precision-Recall
PSY	Psychosomatic Disorders
RD	Rare Diseases
RF	Random Forests
RMSE	Root Mean Squared Error
ROC	Receiver Operating Characteristic
SVM	Support Vector Machine
TN	True Negative
TP	True Positive
TPR	True Positive Rate
UCI	University of California Irvine

Chapter 1

Introduction

1.1 Study Background and Motivation

Rare diseases (RDs), by definition, only affect a minority of people. In the European Union, an RD disease is defined as a disease in which the number of patients does not exceed 5 per 10,000 persons [218]. The majority of RDs are genetically determined, and the prevalence rates of RDs are low [231]. Furthermore, RDs have regional variations and heterogeneity, but the population of RD patients is small and scattered [178]. However, for a population of 7.8 billion, the number of people affected by one type of the RDs may range from hundreds to 400,000 [158]. A conservative evidence-based estimation suggests that 3.5% to 5.9% of the global population is affected by orphan diseases, which equates to 263 – 446 million people [264]. In the European Union alone, about 30 million people suffer from RDs [74]. Orphan diseases develop chronically, progressively, and exhaustively, and they are unlikely to be prevented or cured. The undetected consequences of RDs are often life-threatening, including irreversible health detriments and early death [12].

At present, 7,000 distinct RDs are registered, but only about 1% of them have corresponding medications [178]. Approximately, 5,000 RDs have no epidemiological treatment strategy [231], and some share several common traits [156]. Countless RD patients and families traverse complex medical systems, seeking faster and more adequate treatment at the earliest stage of a scarce affliction. But finding an expert with a matching specialty is often a matter of luck rather than systematic planning by the hospital or national healthcare system. The World Health Organization frequently provides updates on an increasing number of patients receiving delayed or incorrectly derived medical diagnoses from physicians [195, 196]. According to recent statistics, the mean time from symptom onset to a precise diagnosis is around 4.8 years for RDs. On average, patients with RDs visit 7.3 physicians before receiving a correct diagnosis [30, 69]. Research from the European Organization for Rare Diseases (EURORDIS) reported that 40% of RD patients were initially misdiagnosed, and 25% of patients waited 5 to 30 years from the first symptoms appearing to a confirmatory diagnosis of their diseases [74]. More than 20% of patients in the U.S. who seek medical advice receive inadequate or late treatment from medical institutions due to under-measured and understudied symptoms. Sometimes it requires experts to maintain an examination process over months or years to fully understand the disease condition and interpret the different symptom stages that lead to a particular orphan disease [12, 231].

Medical doctors (MDs) at the Medizinische Hochschule Hannover (MHH) have reported on several transfer cases where patients with RDs have had to receive long-term incorrect treatment due to the previous misdiagnosis in other hospitals. However, at the time of diagnosis, patients were usually in the early stages of the disease and with only some symptoms that were not yet significant, thereby leading to misdiagnosis or delayed diagnosis. In particular, the limited information exchange between clinicians and patient feedback management is addressed as a serious impediment for timely diagnosis and treatment. Although reliable tests for the disease in question might be available, lack of relevant knowledge and clinical experience pose challenges to an independent doctor in recognizing and diagnosing the orphan disease. The time from an early indication until the final diagnosis of an RD can take years or even decades [225]. Due to the undiagnosed illness, patients suffer from severe and slowly deteriorating health conditions and invest a significant amount of time and effort on incorrect diagnoses and inappropriate interventions, sometimes even extending to unnecessary surgery. It is undoubtedly a disaster for individual sufferers. The backlog of such cases will impact health insurance and the economy, further influencing the development of healthcare [55].

Committed to ameliorating the diagnosis and treatment of RDs, improving the visibility of orphan diseases, and propagating the benefits of frontier scientific medical research results, Orphanet (www.orpha.net), a coalition of 40 countries funded by the European Commission, established a sizeable platform for orphan disease registration and first-hand information collection and dissemination with a comprehensive population coverage [269]. Over the past 24 years, Orphanet has organized and classified extensive, high-quality information and treatment recommendations related to $6,172$ different RDs with specific clinical conditions, deformity syndromes, morphological and biological abnormalities. The collection of information on RDs has yielded epidemiological data, including health status tracking, disease burden, life expectancy, and healthcare resource utilization standards [264]. The associated orphan drugs and the relationships between RDs, genes, and related symptoms have also been accumulated [216]. The Orphanet database is accessible according to the medical terminology, age of onset, genetic genes, and research resources [264]. A large proportion of RDs with identified clinical practice guidelines (CPGs) have been selected, evaluated, and published in the orphan disease inquiry system [202]. CPGs that meet the audit criteria with their related information are integrated into the Orphanet database, including registered RD patients, medical researchers, disease research institutions, and healthcare professionals who can provide exceptional services from 40 countries.

1.1.1 The Challenges of Diagnosing Rare Diseases

Correctly identifying RDs and determining orphan disease patients' health conditions based on the claimed symptoms is not easy. Clinical signs of an RD must be observed through deductive consultation and inferences about a series of symptoms, while MDs must find adequate treatments to help the patients. However, present research from the medical society states that there exist about $200-300$ apparent symptoms that can refer to $10,000$ known diseases [260]. Different orphan diseases have various characteristics of combinatory appearances. For example, some kidney, liver, and heart problems that appear at a superficial level during childhood are in fact RDs. Even for the same rare disorder, the individual variations among patients will result in different symptoms, and racial and regional differences also impact the characteristics of a disease. RDs are often misdiagnosed if they intersect with other disease symptoms. For instance, osteogenesis imperfecta (OI) might be misdiagnosed as calcium deficiency or a result of polio vaccine [3], and amyotrophic lateral sclerosis (ALS) might be misdiagnosed as cerebral palsy [49, 132, 157]. Furthermore, plenty of RDs with complex causes are still inconclusive and difficult to diagnose because of insufficient knowledge, a shortage of information, and scarcity of domain expertise [264].

The challenge of diagnosing RDs is mainly concentrated on isolating orphan diseases from common, frequently occurring diseases according to patients' symptoms, which often leads to sophisticated processes [35]. The diversity of RDs and their symptoms creates unique difficulties in the diagnostic stage of the disease. Complex pathology and symptoms, missing experiences, and data resources could delay the diagnosis and targeted therapy for RD patients. Statistics show that when multiple influencing factors must be taken into account, overlooked or inaccurately interpreted symptoms of an RD are not uncommon [158]. Doctors' determinations or decision-making abilities strongly depend on the learning effects gained from specific knowledge, previous patients' treatment outcomes, and shared experiences with colleagues or experts from the medical community. The potentially interlinked symptoms causing alternative afflictions must be completely ruled out, which requires MDs to have extensive knowledge about interpreting signs that differ between various types of diseases. However, due to the economic, population, technological, and social security restrictions, there remain some difficulties in clarifying the pathogenesis of specific RDs [12]. Even with a preliminary examination of a patient's physical condition and laboratory tests for delineating the range of possible infections, the potential weakness of cognition about the disease itself and its undefined clinical endpoints make it challenging to identify an RD [231].

1.1.2 Research Motivation

In evolution theory, a disease usually tends to have evolutionary ambitions. For instance, antibiotic resistance demonstrates excellent natural selection in that bacteria change according to their environment [189]. Similarly, the interpretation of an RD requires an immense focus on time series, as well as the recognizable geographic or population variability, to provide the most information on the iterative evolution of RD symptoms. According to evidence-based medicine (EBM), medical decisions should be based on the integration of technologies and methods in clinical epidemiology, statistics, informatics, pharmacy, preventive medicine, and other disciplines [228]. By applying the available evidence and scientific research results, and combining them with MDs' professional skills and clinical experience, further consider the patient's clinical manifestations and personal willingness, an optimal therapeutic measure can be formulated for the RD patient [106].

The inducement of this research originates in the frequently reported diagnosis difficulties and misdiagnosis of RDs. Many organizations and institutions are dedicated to analyzing RDs because the research achievements are an essential symbol of the advancement in modern medical technology and genomics. However, scattered research projects, a scarce RD patient population, and the lack of cooperation make integrating the discovered knowledge about RDs challenging. A successful diagnosis and treatment plan is supported by numerous scientific studies and examinations, which are essentially the conclusion of clinical data deep study. The purpose of analyzing RD data is to transform a vast majority of patients' experiences into universally accepted cognition [257].

In order to develop a method that can be easily operated and promoted for improving the diagnostic accuracy of RDs, medical experts in MHH have proposed a strategy that utilizes the unique characteristics of RD for generating a symptom-associated questionnaire survey. The questionnaire can be applied to facilitate a preliminary screening of RDs from common diseases. For instance, it can determine whether the patient's symptoms are characteristic of an RD. Moreover, the classification approaches with targeted questionnaires for investigating specific subsets of RDs are addressed according to their symptom characteristics. With the idea of the questionnaire-based RD screening system, an interdisciplinary research project that focuses on a computer-aided diagnostic support system was established in cooperate with MDs from MHH and data scientists from Technische Universität Braunschweig (TU-Braunschweig).

The motivation for this study was to perform pattern recognition for RD through statistical analysis of symptom-related questionnaire data, in order to assist in the accurate diagnosis of future patients. The patient's symptoms and perceptions of RD have instructive analytical value. The questions designed by MHH medical experts involve the symptom characteristics of specific RDs, the difficulties that the patients faced in their early stages of RD, and the recorded detours and experiences they have gone through during their diagnostic journey. On the basis of Orphanet, which is devoted to constructing the RD-related information acquisition and inquiry system, it is imperative to establish a collaborative rare disorder diagnostic support system based on data mining and machine learning. By assessing the collected patient symptom data and applying statistical approaches to understand the characteristics of RD, the system can provide a reliable RD screening mechanism, exploit the potential of medical data, and provide sufficient evidence for doctors to understand RD and assist in diagnosis.

The primary demand for a learning and interpreting mechanism is centered on its intelligent statistical analysis capabilities to summarize and refine data characteristics at different stages. With the extensive utilization of data analysis technologies, a computer-aided diagnostic support system, which combines the particularity and complexity of biomedical information, provides a practical approach to presenting information and conclusions through human–computer interaction. Data mining technology has brought high performance, multiple functions, and intellectualization into the fields of disease diagnosis and prediction, disease-related factor analysis, pattern recognition, and medical image analysis [192]. From the research perspective, the analysis of RD-oriented questionnaire data can provide inspiration and valuable ideas for medical research. Collecting adequate RD cases and applying comprehensive analysis can determine features that used to distinguish RDs from common diseases. Solid data support is the foundation of developing advanced disease classification mechanisms and establishing disorder pedigrees, as well as the clinical exploration of targeted therapies and research on the etiology of RDs. From the medical perspective, the computer-aided diagnostic assistant increases the doctors' sensitivity to patients with orphan diseases by screening patients' incipient symptoms and matching RD characteristic patterns. The systematic data analysis narrows the scope of probable diagnosis and avoids the expensive costs of genetic testing. The collaborative study helps doctors and patients gain a sharper insight into rare disorders' symptoms and development trends, benefits diagnosis and treatment, and promotes biomedical research and development in medicines. Indispensable evidence is organized to help decision-makers conduct thoughtful decisions in complex and diversified environments. Doctors' judgments no longer rely on individual experience, but with traceable data support to reduce the misdiagnosis rate and improve the accuracy of diagnosing RDs.

1.2 Research Questions and Objectives

An RD has a low incidence rate and complex pathology, and its symptoms are not always significant. Insufficient understanding coupled with limited pathological evidence and clinical experiences makes obtaining an accurate diagnosis complicated and cumbersome. For medical researchers, the critical challenge is to distinguish the symptom differences between an orphan disease and a chronic disease efficiently in order to improve the RD diagnostic accuracy. Decision support that relies on machine learning (ML) techniques is growing the demand for modern medical research in recent years. Using ML for data analysis can be viewed as pattern recognition or, in a more farsighted manner, knowledge discovery [187]. One category of ML algorithms is supervised learning, which resolves classification or regression problems and effectuates future prediction through data characteristic analysis. Supervised learning methods can help verify hypotheses or assumptions and classify new experimental cases within the predetermined decision ranges [34]. Aimed at realizing a credible method to predict potential patients by utilizing the symptoms-related questionnaire answer patterns of patients who have been diagnosed with RDs, this section addresses the significant investigations in this study and the research objective statements.

1.2.1 Research Questions

In the data set for measuring disease symptoms, the indicators for particular diseases have more or less complex characteristics, and their influence may consist of multidimensional interdependencies. Although a disease's representative characteristics are more likely to be captured by the system, not all of the early symptoms of RDs are particular or typical. On the contrary, they are more similar to the symptoms of common diseases. Due to the complexity of data dependencies in multidimensional symptom associations with various types of diseases, conventional analysis strategies sometimes only provide an insufficient characterization of data. With accumulated medical data (e.g., data concerning various kinds of diseases with their associated symptoms), a trained ML model can interpret potential patterns to distinguish diagnostic classes with their associated symptoms. In the classification problem, supervised ML methods provide pervasive support for exploring data and building predictive models based on experience data. Identified disease patterns can be represented as statistical interpretation models, which are used to decide whether a new patient with symptoms shows characteristics similar to a pre-determined disease pattern. With the ambition to establish a maximally adequate medical diagnosis for each patient, such a statistical limitation with potential risks for interpretation gaps requires a detailed investigation, which is specified as the first research question $RQ1$.

Research Question 1

How can empirically constructed questionnaire data be used to interpret rare disease characteristics and assist future patients' diagnostics?

The identified combination of disease characteristics can be depicted as a statistical interpretation pattern used to determine whether a new patient exhibits the symptom pattern of the identified disease. Advanced statistical analyses distinguish patterns in available data resources and derive accurate predictions based on present and past observations. In many distinct medical areas, there are various successful applications of ML methods, which use past patients' experiences during their diagnosis and treatment journey to predict the current patient's state and future trends [155]. Generalizing and observing patients' symptom patterns of a disease helps doctors comprehend the underlying relationships rather than explain the cause of each symptom.

The process of decision-making is meant to make inferences by continuously expanding the storage of domain expertise and prior knowledge to train and optimize the learning model [155]. A correct decision depends on adequate knowledge representation methods and reasoning mechanisms. In routine disease inspection and interrogation, the standard quantitative questionnaire and answer process determines the initial diagnosis of a disease by referencing the closest disease pattern to a new patient answer sequence. However, in the case of RDs, the complex correlation between symptoms is not always intuitively explained. In addition, the early signs of RDs sometimes exhibit similar characteristics to ordinary diseases. The prominence of a single symptom may lead to misdiagnosis.

Although sophisticated statistical analysis systems identify the symptom dependencies and idiosyncratic answer patterns, correlation does not represent absolute causality. The patient's final diagnosis still requires domain experts to combine system prediction with a series of medical examinations to reach a comprehensive conclusion. MDs, who are the targeted users of the computer-aided diagnostic support system, must understand and master the methods of extracting and reviewing classification results. Medical practitioners can expand their disease identification ability by evaluating a system diagnosis's effectiveness while improving system performance by calibrating false diagnoses in practice.

Nevertheless, not all doctors are necessarily familiar with modern computational analysis mechanisms. There is an invisible barrier between data modeling and knowledge extraction. Classification models can be developed through diversified ML algorithms. If the results expressed by the model are incomprehensible and users cannot understand why the model makes specific decisions, the performance of the model will become powerless. Human cognitive factors must be considered in the process of knowledge extraction. The explainability of a model, especially the complex nonlinear model, is a decisive factor in applying ML-based decision support to clinical practice. Otherwise, the black-box nature of the ML model may significantly impede its widespread adoption in the medical field [155]. To facilitate a transparent understanding of the system prediction path from data analysis to decision-making, the second research question $RQ2$ is as follows.

Research Question 2

How can doctors be logically assisted with different knowledge reserves to understand the correlation between distinctive rare disease symptoms and computer-aided diagnostic results?

In most cases, data visualization and imaging support can help scholars who have not accumulated enough experience to better understand the data analysis procedure. This is because the human visual system is more sensitive to images and colors than text. In other words, digital imagery allows users to acquire knowledge and capture critical information more intuitively. When appropriate visualization techniques are applied to the characteristic recognition and analysis results of the research object, the interpretability of the numerical results will be enhanced because of the additional accessibility in cognitive perception [262].

Considering the fact that this study focuses on the analysis and classification of RD-related questionnaire data, the visualization approach boosts the discovery of disease characteristics. In particular, it highlights the essential traits of a disease category while excavating potential knowledge. Data visualization expands the advantages of pattern recognition technology for diagnostic support. It can change people's misgivings about which questions or early symptoms are most worthwhile to distinguish and which combination of questions can lead to the result more efficiently than others. Through the intuitive observation of question dependency, some particular types of diseases provide recognizable answers structures, resolved from a combination of several distinctive questions. However, in the complex case of RD screening, it is crucial to avoid confusion and noisy data caused by redundant or irrelevant questions to the patient's proper disease. Patients are often held to fill in extensive and time-consuming surveys, which leads to delays in getting patients' responses and invalid information. To avoid respondents getting overwhelmed by spending too much time and effort on answering all questions in a questionnaire, analyzing the correlation between questions and the diagnostic result while finding a way to reduce unnecessary questions was on the agenda.

During the computer's heuristic learning phase, the data's integrity and validity need to be guaranteed to provide a robust foundation for supporting different hypotheses and classification models. Well-trained classification models can summarize a stabilized answer pattern under each category to provide strong evidence for system diagnosis. When examining new patients by collecting their questionnaire responses, patients themselves would not recognize which questions are related to their potential disease and which are not. The irrelevant questions would interfere with patients' responses and negatively affect computer-aided systems' diagnostic accuracy due to invalid or erroneous data acquisition. In view of the above considerations, the third research question $RQ3$ is defined as follows.

Research Question 3

How can patients' questionnaire complexity about rare disease conditions be simplified while maintaining the accuracy of computer-aided diagnostics?

Under these circumstances, an dynamic adaptive questioning mechanism with a more channelized focus would be highly desirable to investigate. It would reduce the questions to only the most relevant ones for the target individual, therefore shortening the elapsed time substantially and increasing efficiency. The symptom dependency analysis and visualization of the RDs patients' characteristic data paved the way for the further improvement of the questionnaire, especially the development of a user-oriented questioning procedure. The substance of $RQ3$ focuses on the exploration of appropriate strategies that determine which symptom combination can best distinguish different types of diseases. Furthermore, it seeks additional clarity in comprehending the disease characteristics and making classifiers achieve more accurate predictions.

1.2.2 Research Objectives

This exploratory research aims to realize a classification mechanism based on ML methods and pattern recognition to provide RD detection and diagnostics at the early stage of illness. Patients' symptoms and experiences of orphan diseases during the diagnostic journey have a high analytical value, as they have the potential to help MDs gain experience and prepare for future RD diagnoses. The primary focus is whether the early symptoms of RDs can be investigated and interpreted by questionnaire data that patients provide. By collecting disease-related empirical data from diagnosed patients, medical experts from MHH designed and released a questionnaire in order to isolate RDs from common diseases. Furthermore, a series of questionnaires were designed to identify some specific RD categories, such as neuromuscular diseases and pediatric primary immunodeficiency disorder (PID). With the in-depth exploration of research questions, three research objectives were determined, as shown in Table 1.1.

Table 1.1: Overview of Research Objectives

	Objectives	Methods	Related RQs
1	To develop a diagnostic mechanism with well performed classification method to identify rare diseases	• Statistical Analysis • Machine Learning Algorithms	*RQ1*
2	To provide a platform to explore the underlying rules and principles of rare diseases through emblematic symptom patterns	• Statistical Diagrams	*RQ2*
3	To formulate a questioning mechanism for predicting possible disease of new patients through channelized collection of most representative symptoms	• Adaptive Mechanism • Pattern Recognition	*RQ3*

The first objective was to develop a reliable diagnostic mechanism to identify RDs based on symptom-related questionnaire data. The method used in this study is based on the fundamental concepts of supervised ML algorithms and statistical analysis approaches. As the concept of $RQ1$, to provide maximum decision support for RD diagnostics, this research utilized the strengths of different ML algorithms and consolidated a well-performed classifier fusion method. In diagnosing a new patient, the diagnostic mechanism analyzes the qualitative and quantitative relationship between the incoming questionnaire data and the characteristic pattern of RDs. By seeking the most-matched symptom condition, the system rapidly and accurately proposes the diagnostic result.

For answering $RQ2$, the second research target was to provide a platform that can aggregate each RD's distinctive symptoms and visualize the determinants of possible diagnoses. The development of visualization functions was aimed to interpret the disease characteristics intuitively, thereby helping MDs understand the inferences made by the diagnostic mechanism and explore the intrinsic principles of RDs through emblematic symptom patterns. The supplementary involvement of a visualization module can increase the explainability and interactivity of the system predictions with the joint efforts of medical specialists, statisticians, and data analysts. The combination of statistical analysis and visual representation simplifies the interpretability of the classification results, without the demands of experts' commentary or additional sophisticated background knowledge.

As the adequate concept for $RQ3$, the third goal was to formulate an adaptive questioning mechanism based on classification to predict the possible disease of new patients through a channelized collection of representative symptoms. In a static questionnaire, some seemingly insignificant questions may have a small impact on the final diagnosis. Still, the combination of multiple questions will influence the disease interpretation and orient the system to make a definitive diagnosis. Questions that have less relevance to the disease might become interference factors for decision-making. The elimination of irrelevant variables and reduction of noise information helps the diagnostic support system improve the investigation efficiency. By analyzing patients' answer patterns and the relationship between variables, raising customized questions based on each patient's response to previous questions enhanced the diagnostic support system's performance.

1.3 Conceptual Architecture Design

The questionnaire-based data analysis and RD diagnostic support methodology were designed to help MDs integrate the RD patients' symptoms and possess the disease pattern information to facilitate diagnosis and treatment for future patients. With the purpose of constructing a unified RD analysis platform, this dissertation endeavors to research a comprehensive disease diagnosis improvement method that involves questionnaire data preprocessing, multiple ML models training and evaluation, disease clinical manifestations aggregation and pattern recognition, and adaptive online disease prediction.

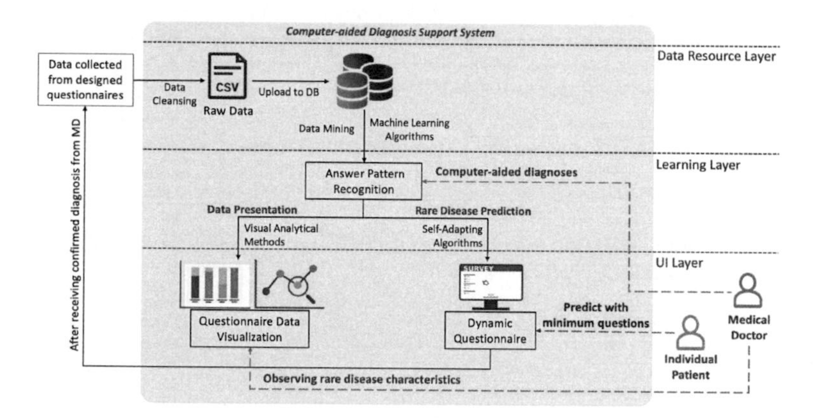

Figure 1.1: Conceptual Architecture: Computer-Aided Diagnosis Support System

By investigating the research objectives and functional requirements, a three-layer system with the composition of an interactive visual interface, an integrated ML model, and information resources was constructed. Figure 1.1 presents the conceptual design concerning the architecture of the computer-aided diagnostic support system. The adequate support for the system's major theoretical and functional achievements relies on three dominant layers.

The main obstacle to diagnosing RDs in a timely and effective manner is the insufficient combination of theoretical cognition with practical explanation and information support. For medical researchers, whether their central focus is on analyzing the mechanism, etiology, disorder pedigrees, or exploration of clinical therapies on targeted RDs, the study required sufficient qualified data support [122, 257]. In order to meet the purpose of information utilization, a proper data cleansing and standardization process were indispensable to ensuring the accessibility, validity, preciseness, and integrality of the data resource. The systematic framework requires substantial information support concerning the diagnosis through disease symptom investigation. As the infrastructure of a decision system, the data resource layer delivers valuable data that consist of patients' essential physical information, responses to questions, and confirmed diagnoses of related RDs.

In the data acquisition stage, medical experts in MHH objectively collected valid RD patients' data from clinical observations, patient interviews, and questionnaire investigations. Appendix A describes a detailed devisal of the disease symptoms-based medical questionnaire and the data collection methods. Each question in a medical questionnaire is designed to measure a certain peculiarity of the concerned rare disorder. The collected questionnaire data provided a solid foundation for extracting useful information and understanding the correlations between diseases and symptoms.

Before the system applied collected patients' responses for data mining and classification model training, a data cleansing process was applied to fix errors in the raw data set (e.g., duplicated, incomplete, incorrectly formatted data, or irrelevant to the objective of the data set). The disease classification and prediction model could make a reliable diagnosis based on the exhaustive study of accumulated disease characteristics and the response pattern of diagnosed patients. In the computer-aided diagnostic support system, a database is aimed to achieve centralized management of collected questionnaire records from diagnosed patients. The goal of the structured database is to ensure data consistency, reasonable redundancy, data storage standardization, and maintainability. For promoting the ML-based diagnosis support system to operate effectively and synchronized with the latest collected empirical data in time, an authenticated MD can upload valid data records corresponding to the registered disease questionnaire as the occasion demands. Detailed data cleansing process and storage structure are introduced in Appendix B.

Integrating suitable data mining methods and appropriate knowledge discovery applications is essential to providing rapid function invocations to explore the maximum value of the questionnaire data. In the learning layer, the system deduces arguments and conclusions through data mining and ML algorithms. The diagnosis support system assembled multiple classification models as the kernel for disease pattern recognition, and each prediction model was trained with the diagnosed patients' questionnaire data (see Chapter 3). Answer patterns for specific RD types were generalized by analyzing the responses of disease-related patients. Summarizing diagnosed patients' exhibitions can distinguish symptom differences between orphan diseases and ascertain the most valuable questions to diagnose certain disorders. Subsequently, a recommendation mechanism for MDs was embedded to elicit system predictions by matching new patients' symptom characteristics (acquired by questionnaire) with the aforementioned answer patterns of different diseases. The probability score provided by the classification model reflects the potential impact of latent disease and possible future conditions. ML demonstrated its power with accumulated historical data and provided sufficient evidence to assist decision-makers with thoughtful predictions in complex and diversified environments.

An interpretation platform is needed to improve the classification model's comprehensibility and exhibit the validity and reliability of the prediction results. According to the results of data exploration, two interactive functions were provided through the UI layer. As an exhibition window, Chapter 4 introduces a graphical platform that provides visualization support for different disease symptom patterns and presents the characteristics of RDs in a human-understandable form. A mapping function was implemented to intuitively compare individuals' questionnaire responses with the disease patterns or discover gaps in symptoms between different diseases. The data visualization module improves the MD's comprehension of the relationship between disease symptoms and diagnosis results through visual transmission and perception.

By analyzing the determinants of classification results, the question variables associated with a particular disease provided the basis for implementing the dynamic adaptive questionnaires. A question significance ranking table concerning the variable correlations and the influence of a single question in the classification process can be found in Appendix C. Such dependency measurement can be applied to the self-adaptive question pushing mechanism introduced in the individual-oriented dynamic questioning method (see Chapter 5). The adaptive question assignment based on real-time diagnosis provides channelized and customized questionnaires consistent with patients' actual conditions. This approach can eliminate irrelevant data, increase the questionnaire compactness, recognize patients' diseases or traits with minimum responses, and improve system prediction efficiency. What is more, in order to collect reference data extensively, once a new patient obtains a confirmed diagnosis from a MD, their questionnaire record will be collected as the foundation for future analysis and classification model training.

1.4 Research Achievements and Contribution

The achievements of statistical analysis in medical domains require integrating different data types, such as clinical, genetic, and biochemical data, with supplementable computational tasks (including descriptive or predictive purpose), model design constructs, and appropriate learning techniques. In this research, qualified questionnaire data were collected from diagnosed patients who were subordinated to their predetermined disease group. A computer-aided diagnostic support system was progressively designed to assist physicians and medical researchers during this doctoral study period. More than eight associated papers were published internationally in reputable data analytics and medical diagnosis conferences.

1.4.1 Accomplished Publications

Concerning the first research objective, a supervised ML task that employs an ensemble of diversified classifiers with a sample-based, weighted voting scheme was established. The ensemble method benefits from the individually functional advantages of each classifier by evaluating their compatibility and accuracy. The paper with the title "Diagnosis Support for Orphan Diseases: A Case Study using a Classifier Fusion Method" [140], published in the ***Seventeenth International Conference on Intelligent Data Engineering and Automated Learning (IDEAL2016)***, introduces the classification core of the computer-aided diagnosis support system. The functionality kernel builds on a fusion approach that combines four classifiers with outstanding performance on the collected orphan disease questionnaire data set. The selected classifiers are support vector machines (SVM), linear discriminate analysis (LDA), random forests(RF), and logistic regression (LR). Each classifier was trained with questionnaire records from previously diagnosed patients and predicts new patient cases independently. When it comes to diagnosing the new patient, a set of prediction probability scores was provided by different estimators, where each probability score indicated the level of confidence in classifying the sample to a specific disease group. By averaging the probability scores from the aggregated learning methods corresponding to each disease category, the ultimate diagnostic label was assigned by the disease type that gained the highest probability.

Subsequently, a system optimization relies on probability calibration, and refinement case-based decision-making was proposed in the paper "Improving the Decision Support in Diagnostic System using Classifier Probability Calibration" [141], which was published in the ***Nineteenth International Conference on Intelligent Data Engineering and Automated Learning (IDEAL2018)***. A calibrated probability comparison table was generated for a follow-up screening process after the classifier fusion method. The table was used to retrieve K neighbor cases from the recorded questionnaire dataset according to the calibrated prediction probability of the new patient. By extracting profitably evidence to support the system prediction result and discuss different scenarios according to the case similarity ratio, the system was able to examine the reliability of the diagnostic proposal, thereby giving appropriate recommendations.

The theoretical and practical study results confirm that abundant investigation records from patients with an affirmed diagnosis can provide sufficient evidence for statistical analysis and disease pattern recognition. The rational use of collected data provides evidence for computer-aided interpretation and diagnosis support of new patients with the stated symptoms. Furthermore, the forecast accuracy is closely related to the performance of selected classifiers. The experimental evaluation showed that the ensemble of classification methods gained considerable disease prediction accuracy compared to any of the involved individual classifiers. The optimized diagnostic support system provides an improved measure for adequate diagnosis, especially when the patient's symptom indicators are insufficient to clearly determine the correct disease. The flexible and effective prediction process helps interpret RD patients' symptoms and assists MDs in obtaining a more precise diagnosis.

As the research result of the second objective, the paper titled "VoQs: A Web Application for Visualization of Questionnaire Surveys" [285], published in the ***Fourteenth International Symposium on Intelligent Data Analysis (IDA2015)***, introduces a data visualization tool aimed at helping researchers collect important attributes, grouping questions, and observe the answer pattern of different categories. It supplements the medical data interpretation capabilities for nonprofessionals in examining the answer probability distribution of each question, seeks the pronounced characteristics to distinguish different disease categories, determines the possible internal relationships between symptoms and the diagnoses, and identifies the feature differences between similar diseases. By showing various associated diagrams, the visualization model offers different observation aspects of the questionnaire data, accentuates essential disease characteristics, and discovers unusual cases that deviate sharply from general answer pattern of specific RD. The user-friendly operating system bundles the associated graphics together to meet the requirement of targeted analysis.

Benefiting from the self-adaptive, self-learning, and self-organizing intelligent computational mechanisms, the paper "A Dynamic Adaptive Questionnaire for Improved Disease Diagnostics" [139] published in the ***Sixteenth International Symposium on Intelligent Data Analysis (IDA2017)*** covers the realization of a dynamic question inquiry strategy. In view of the third research goal, the channelized question pushing method described in this paper enhanced the functionality of the questionnaire system and the performance of the classifier fusion model. In contrast to conventionally conducted surveys with static question sets, the involved questioning mechanism adaptively retained only the most relevant variables to the target user and eliminated redundant questions. The validation results demonstrate that the dynamic dimensionality reduction could gather sufficient information to support or contradict a suspected diagnosis tendency. Meanwhile, it increases the accuracy of the diagnostic support system by alleviating noise data influence. The study indicates that the adaptive questionnaire could help scientists, researchers, and MDs reveal patients' symptom information and reduce unnecessary computation costs. It also enables extended support for the interpretability and classification of associated symptoms for specific RDs.

1.4.2 Contribution to Medical Research

In practical applications, the implementation of the computer-aided diagnosis support system based on ML methods has made significant contributions to several medical research projects. In the medical investigation through Germany about the necessity of analytical support for RDs and shared pre-diagnostic phenomena, Blöß et al. [35] proposed a crucial need for an effective countermeasure to improve diagnostic accuracy and reduce the RD misdiagnosis rate. The qualitative and quantitative analysis results indicate that a predictive system that prompts general practitioners of possible RDs and recommends further clinical examinations when patients present symptoms consistent with RDs is urgently needed. Medical experts hypothesized that detecting inherent patterns from the acquisition of shared characteristics and experiences among diagnosed RD patients could indicate an RD's presence. By combining relevant data analysis results and experimental conclusions, the research found that the strategy of ML-based disease pattern recognition could assist doctors in observing and summarizing the common symptoms of RD patients, therefore optimizing the doctors' capacity to discern RDs. The development and implementation of the diagnostic support application have the potential to reduce the diagnostic latency and the rate of inaccurate diagnosis, thereby facilitating the diagnosis process of individuals and benefiting future RD patients.

The classifier fusion method that combines six supervised ML algorithms was applied to identify questionnaire response patterns of specified neuromuscular diseases [92], and a weighted majority voting strategy was used to propose the final diagnosis. There were collected 274 valid questionnaire records covering seven diseases. Among them, 210 individuals' questionnaire records were used as training data for 21-fold cross-validation. The rest of the 64 questionnaire data formed the test set. The validation result shows that the classifier ensemble method achieves an average diagnosis accuracy of 89.5%. In the prospective test phase, 89% of the questionnaire records were correctly classified. Remarkably, among the 64 test data, nine patients who the classification model diagnosed as having RDs did not receive a precise MD's diagnosis when submitting their questionnaire responses. Revisiting results showed that the system predicted 7/9 of patients correctly. The receiver operating characteristic (ROC) curve and p-value analysis confirmed that fusing multiple well-trained classifiers enhanced the diagnosis accuracy more than a single classifier. The experimental result demonstrated that the classification model has a robust performance in predicting neuromuscular diseases.

In the medical research that focuses on analyzing patients' experiences with pediatric PID [185], the embryonic form of the computer-aided diagnostic support system was applied to assist pediatricians in identifying suspicious PID patients and improving clinical diagnostic accuracy. Medical experts found that the early clinicopathologic features provide clues for potential RDs. The experimental evaluation of the diagnostic support system was divided into two phases. In the first phase, in which 64 PID patients and 35 healthy children participated, the 11-fold cross-validation indicated that 98% (97out of 99) of individuals received the correct diagnosis. In the second stage, which additionally involved 27 children infected by other diseases, the 21-fold cross-validation revealed that 90% (113 out of 126) of questionnaires could be predicted correctly. The data visualization module was used to observe the answer pattern differences between PID patients and non-PID individuals. The research results indicated that ML-based answer pattern recognition performs well in identifying patients with PID. In addition, the data visualization function can efficiently present questions with high significance and the potential correlations between symptoms.

Sieg et al. [238] utilized the computer-aided diagnostic support system to identify patients' response patterns and disease characteristics in the research that emphasized the diagnosis of rare metabolic diseases. The validation results demonstrated that the proposed system can correctly recognize 90% of questionnaire responses from the test data set. The medical research results indicated that the classifier fusion method has a more significant performance improvement compared with a single classifier. Moreover, the data visualization procedure involving Fisher's exact test can distinguish decisive factors in system diagnostics. The predictive diagnostic support approach has the potential to support RD screening, and it can help MDs gain insight into the implicit information and detect individuals with rare lysosomal storage disorders in the future. In the research on common pre-diagnostic features of RDs, Grigull et al.[93] applied the computer-aided diagnostic support system to distinguish the questionnaire answer characteristics of patients affected by RDs. The cross-validation results verified the flexibility and malleability of the diagnosis support system. They also revealed that the diagnostic support system could identify 88.9% of RD patients. The embedded classification model can efficiently recognize different answer patterns, and the visualization function explicitly expressed the diagnosis evidence. Utilizing the experience of diagnosed patients as references for identifying future patients is effective because although the manifestations and pathogenesis of each RD are very different, surprisingly, patients with RDs share similar pre-diagnostic experiences.

These medical studies substantiate that the strategy of combining different but complementary ML algorithms leads to sufficient performance in pattern recognition and disease prediction. Systematic analysis of disease symptoms and characteristic visualization have brought convenience to medical researchers and effectively improved general practitioners' insight of RDs. Since the diagnostic support tool operates independently of physicians' experiences, it can cover the shortage of RD expertise, derive recommendations for further actions, and shorten the path to a definitive diagnosis. The ML-based diagnostic support system has high popularizing value and can be continuously improved during data accumulation. Appendix D enumerates the contribution of the ML-based diagnostic support system to several medical projects in the academic field.

1.5 Structure of the Thesis

In the medical domain, data analysis makes it easier to grasp disease characteristics, symptom iterations over time, study ethnic or regional differences based on relevant references, and draw conclusions that can be disseminated as knowledge. With the support of data analysis, doctors are able to make more reliable decisions and appropriate interventions in a timely and moderate manner. However, exploring the RD-related data is a challenging task where the requirements for data-driven methodologies are continually increasing — for example, efficiently gathering high-dimensional data from designed questionnaires, clinical trials, and research experiments, facilitating the comprehensive utilization of data samples from a particular population, and converting data information into knowledge and cognition. Decision support systems that classify objects into different categories are based on the feature descriptors that fall under the rubrics of ML algorithms. In the questionnaire-based diagnostic support research project, significant progress was made in data collection, pattern recognition, feature extraction and association, and disease prediction. Significantly, this research optimized the process of orphan disease determination and developed an interactive-oriented integration framework, which fulfills the gap of RD observation, interpretation, and ratiocination based on ML methodologies.

In this thesis, exhaustive research was conducted regarding data acquisition and processing, data analysis technologies, and data presentation methods. Furthermore, the performance of data analysis methods in the prediction and prognosis of medical disease was investigated. Chapter 1 introduces the project background and research motivation, and enumerates the importance, necessity, and benefits of this research. This chapter also determined the precise research questions and primary objectives. A conceptual system architecture design was proposed based on a detailed analysis of the major functional requirements. Based on progressively accumulated research results, the resolution strategies for each research objective are proposed through the brief introduction of the published papers during the doctoral study period. Section 1.4.1 contains four relevant papers published in pioneering data analysis conferences. Two papers are dedicated to the construction design of the disease classification model, research on improvement approaches to enhance the computer-aided diagnostic accuracy, and the validation of the diagnostic results. One article introduces visual representations of the questionnaire data and corresponding formulas for data integration and transformation. Another paper proposed the process of dynamic adaptive questionnaire strategy, and the effectiveness of the self-adapting procedure was verified with real-world data. Section 1.4.2 explores the implications of this research on the clinical practice of disease diagnosis and academic contributions. The achievements of this doctoral research received recognition and acceptance from MHH medical experts.

On the basis of computational statistics and mathematics, Chapter 2 covers the statistical learning and data analysis theories that are mainly involved in the research of RD diagnostic support, including data mining methods, ML techniques, classifier ensemble, model validation and evaluation, and data visualization methods. Typical ML applications in the area of medical data analysis and bioinformatics are reviewed, and their advantages and disadvantages are discussed. Chapter 3 introduces the kernel of the designed RD diagnostic support system, which covers the explanation of the calibrated classification model. A classifier ensemble method and isotonic regression is proposed to optimize the learning and pattern recognition process in order to maximize the accuracy of disease prediction. Chapter 4 focuses on the implementation of an interactive visualization platform. By observing patients' questionnaire data from different perspectives, the visualization approach helps MDs better understand RD characteristics and diagnostic bases. Data visualization improves the interpretability of diagnostic results and provides inspiration for further exploration.

In Chapter 5, a dynamic adaptive question routing mechanism is introduced to select questions that efficiently describe the target patient's symptoms. The system releases further questions based on previous questions' feedback or the combination of earlier responses given by the patient. The validation results regarding the overall accuracy of the diagnosis support system indicate that the dynamic questioning mechanism can effectively reduce the complexity of the questionnaire while improving the efficiency of data collection and the capability of diseases' symptom interpretation. Under the premise of guaranteeing the quality of prediction results, this method significantly reduces noise or irrelevant variables by dynamically seeking the correlation between questions and the final diagnostic in the interactive system. The system's diagnostic accuracy can be used as an indicator for assessing whether the intelligent questioning mechanism and the classifier fusion method benefit MDs and patients to increase the opportunities of receiving an early diagnosis and adequate treatment.

A comprehensive conclusion of experiences and knowledge gained during this research period is described in Chapter 6. It discusses the construction and achievements of this doctoral research by addressing the answers to three research questions, together with the limitations encountered in this study, acquired skills and abilities, and the research prospects and future directions of the computer-aided diagnostic support system. The intervention of computer-aided decision support has optimized the process of disease diagnosis in the medical field, particularly improved the diagnostic accuracy of RDs, and strengthened the medical researchers' perceptiveness of the symptoms of orphan disorders. The results of this doctoral research not only assist in the diagnosis of intractable diseases and significantly reduces the patient's consultation cycle, but also pave the foundation for further theoretical research. More importantly, the ideas and methods mentioned in this thesis can be popularized and extended to other fields.

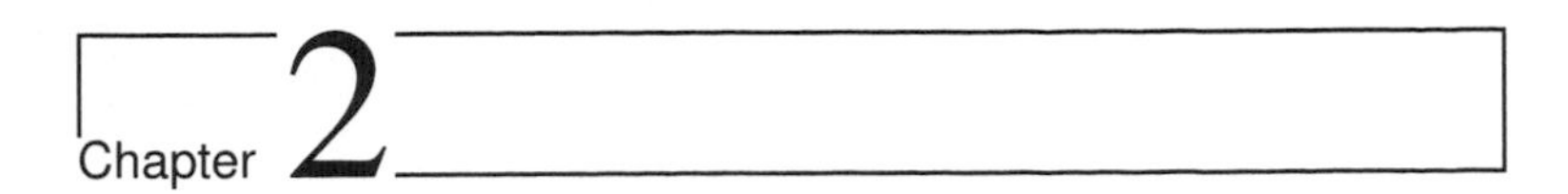

Fundamentals and Related Work

The value of data lies in examining and discovering the essence of phenomena. Appropriate data processing and statistical analysis can extract useful information from the data source that is conducive for information holders with different analytical backgrounds to distinguishing and understanding related data, summarizing knowledge to make appropriate decisions, and developing strategies. Since RD diagnosis requirements are rather complex and the knowledge base of RDs is in the process of continuous exploration and expansion, a single reasoning technique may not always be reliable or universally applicable (the "no free lunch" theorem [275]). This chapter reviews and discusses the theoretical methods involved in researching questionnaire-based RD diagnostic support and typical applications in the medical research area, covering standard ML algorithms, model evaluation methods, fusion methods, and the implementation of data visualization. As an assistive tool for medical diagnosis, ML algorithms learn and summarize the characteristics of known samples and form a feature pattern to identify unknown cases. Integrating multiple learning models is a practical approach to improving a system's predictive ability. The combination of accurate and diversified learners can improve the final predictability with their strengths. Appropriate data analysis and visualization methods help transform information into valuable insights and present clear conclusions in graphical form.

2.1 Statistics in Diagnostic Medicine

Knowledge discovery in databases (KDD) is an interactive, repetitive process of extracting valuable information from large amounts of data [79, 80]. With the widespread application of statistical analysis methods and data acquisition techniques, many scientific research projects in the field of life sciences have transformed their traditional research models to computer-aided data analysis methodologies [75, 239]. The primary purpose of utilizing statistical thoughts and data mining methods is to concentrate information from various data sources and find relationships between different variables [80, 271]. The statistical analysis techniques help extract information from data and build the probabilistic learning model to discover underlying phenomena and hidden knowledge [99, 243]. The study of historical data helps identify decision rules and predict future trends with minimal manual intervention [9, 200].

In medical science and healthcare research, inference and prediction are the major concerns of statistical analysis projects [75, 239]. The ultimate goal of studying medical data is to assist medical practitioners in screening through the collection of possible diseases, providing reasonable diagnostic recommendations, and proposing appropriate therapeutic schedules [82, 138]. Implementing a computer-aided decision support system with efficient knowledge discovery methods can help analyze the complex relationships from clinical data and extract potential medical cognition [26, 91]. The computer-aided diagnostic systems recognize the symptom patterns of different diseases using multidisciplinary knowledge, such as statistical, probabilistic, and optimization technologies [199]. By revealing the potential inevitability and possible regulation from the seemingly complex contingency, statistical analysis algorithms assist specialists in analyzing the mechanisms of the diseases, gaining insight into the essence of medical problems, and providing sufficient data support for doctors' verdicts [18, 200].

As a branch of artificial intelligence (AI), ML methods concentrate on developing automated data analysis models that can recognize intricate patterns and intelligently make decisions based on large, noisy, and complex data [79, 98, 187]. By analyzing and representing the correlation between various symptom indicators and specific diseases, the ML algorithms can learn the rules of distinguishing particular disorders from the diagnosed patients' clinical data and discover disease patterns [26, 114]. The published studies and experimental findings substantiate that the ML-based disease pattern recognition can programmatically summarize the symptom differentiation rules, thereby providing a crucial resource for predicting unperceived trends [114]. Ilyasova et al. [115] compared five different statistical approaches in the research of computer-aided eye disease diagnostics. The research results indicate that the discriminant analysis-based algorithm can efficiently improve the feature informativeness for distinguishing intergroup differences and classifying fundus patterns. Cruz and Wishart [59] investigated the performance of different ML methods in cancer prediction and prognosis. The research suggested that well-constructed ML-based diagnosis systems can significantly enhance the accuracy of predicting cancer susceptibility, recurrence, and mortality. Compared to conventional statistical or expert-based systems, ML methods generally improve the performance of disease prognoses [214].

Choosing the appropriate ML methods according to data type and research purpose can maximize the value of medical data and enhance the overall quality, generalizability, and maintainability of ML-based classification systems [23, 166, 167]. Within artificial intelligence, there are mainly three ML approaches: supervised learning, unsupervised learning, and reinforcement learning. Figure 2.1 presents the structural differences of these three learning paradigms [173].

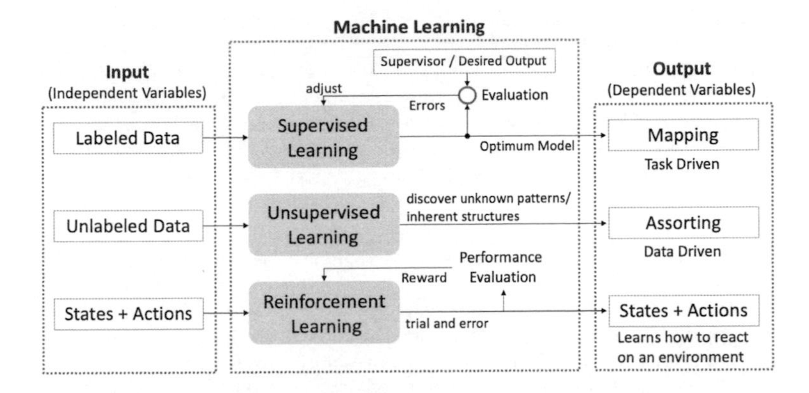

Figure 2.1: Comparison of Machine Learning Paradigms: Supervised, Unsupervised, and Reinforcement Learning [173]

In supervised learning, each data record is marked or associated with a category or value of interest (labels). Supervised learning tasks can be further divided into classification problems and regression problems. Classification focuses on the problems that whether the predicted variables are discrete, while regression focuses on issues that whether the predicted variables are continuous [81]. The goal of supervised learning is to study examples with such labels, thereby being able to predict the class of future data [64]. Tsumoto and Shusaku [257] used a rule induction method to extract diagnosis rules from experts and generalize diseases' properties from the congenital malformation database. Kuo et al. [153, 154] applied data mining with a trained decision tree model to increase the diagnostic confidence of classifying breast tumors in ultrasonographic feature investigation. By using intelligent selection algorithms to adjust parameters, the research revealed that the proposed ML-based diagnostic system could minimize risks for misdiagnoses and simplify symptom interpretations by learning from accumulated historical data. The computer-aided diagnostic support system significantly enhanced the diagnostic accuracy compared to experienced physicians. Grounded on the combination of generalized discriminant analysis (GDA) and the least-squares support vector machine (LS-SVM), Polat et al. [208] proposed a cascade learning system to distinguish healthy and diabetic patients by characteristic variables. The evaluation results confirmed that the hybrid intelligent diagnostic assistance obtained better classification accuracy and robustness than a single classifier.

On the contrary, there is no label correspondence for data points in the unsupervised learning. Unsupervised learning problems can be divided into association problems and clustering problems [173, 179]. The goal of an unsupervised learning method is to organize the data and model the underlying structure or distribution, grouping data into a cluster to observe complex data in a simple manner [105]. For instance, Marlin et al. [172] developed a probabilistic clustering model to discover patterns in electronic health care data. The research results indicate that the proposed model can find distinct, recognizable physiologic patterns with prognostic significance. Rustempasic et al. [226] introduced a Parkinson's disease diagnostic system using fuzzy c-means clustering and pattern recognition. By training the model with the Parkinson's data set containing human biomedical voice, the experimental results demonstrated that the system can effectively recognize and distinguish the differences in speech signals between normal speakers and Parkinson's disease speakers and automatically detects Parkinson's disease.

Reinforcement learning refers to goal-oriented algorithms [247]. It is a continuous procedure intended to train the system to make a sequence of decisions to attain its objective in an uncertain, potentially complex environment. The reinforcement learning algorithm selects an action among the list of possible activities for the corresponding data sample, learns how to behave in the environment by interacting with it, and receives execution rewards to maximize the cumulative performance. Reinforcement learning will modify the ensuing strategy based on each piece of feedback and iteratively repeat until the model converges to achieve a higher rate of return [13]. Liu et al.[163] integrated representative deep reinforcement learning models with medical data in the computer-aided diagnostic support system to detect common types of lung cancer and summarize their characteristic patterns. Fatemeh Mansourypoor and Shahrokh Asadi [171] developed a reinforcement learning-based evolutionary fuzzy system to build and optimize the diabetes diagnosis rule. The experimental results show that the proposed methods provide a compact and accurate system with higher robustness for diabetes diagnostics.

Mahmud et al. [168] and Yu et al. [281] retrospected and summarized the theoretical basis and the application of deep learning and reinforcement learning techniques in mining biological data and analyzed the performance of decision-making in different healthcare domains. Both articles suggest that reinforcement learning can present a mathematically and technically solid solution to optimize decision-making in dynamic treatment and automated medical diagnostic tasks. However, reinforcement learning has not received as much attention in computer science or medicine compared to traditional ML, statistical algorithms, or theory-based approaches. Since reinforcement learning requires a significant amount of computing power and memory with a relatively slow convergence rate, it is not worth applying to small or medium data sets. In addition, reinforcement learning has poor interpretability, and it is inapplicable to real-time applications due to the inadequate processing capability of experimentally collected data.

The application of ML algorithms in healthcare is interdisciplinary research drawing from computer science and medicine. It is a collaborative process requiring the joint efforts of ML researchers and specialists from different disciplines to use their expertise in their fields to develop appropriate solutions for optimal disease diagnosis and treatment formulations. Musen et al. [186] proposed that the construction of diagnostic support systems brings substantial social benefits by improving medical care and explaining the therapeutic decision-making process. By accumulating knowledge through data mining and expert systems, the ML-based diagnostic support approach assists doctors in identifying patients with specific conditions, risks of complications, or requirements regarding interventions (such as screening tests or immunizations) [243]. With the combination of medical practitioners' experiences, diagnosis principles, and theoretical medical foundations, the computer-aided diagnosis and recognition model can effectively avoid misdiagnosis and identify appropriate treatment suggestions based on observed symptom variables [26, 91, 111, 266]. Compared with conventional statistical or expert systems, the assistance of ML methods improves the precision of disease determination and prediction [8]. Medical practitioners can comprehensively obtain new patients' disease information and corresponding possibilities [23, 205]. Meanwhile, the ML-based system can help doctors integrate experiences and enrich medical theories, as well as the understanding of disease development and progression for clinical exploration and targeted therapies [122, 205].

Since the research subject of this computer-aided diagnosis support system is a challenging RD prediction task, the supervised classification methods perfectly meet the requirements of feature extraction and category determination by analyzing historical cases. Supervised ML uses labeled instances for inference learning, generates and trains diagnostic mechanisms through corresponding algorithms, and then builds an accurate model to predict future examples and discriminate between different classes. For example, logistic regression summarizes features through generalized linear models, support vector machines build classification models through high-dimensional feature transformations, and k-nearest neighbors make predictions based on similar cases. In order to incorporate ML into medical fields that require high precision, there are certain requirements for developing diagnostic aids. Considerable effort is required to provide sufficient clinical cases with the same set of informative features, and the system must be infused with extensive background knowledge in the relevant clinical field. Physicians' ability to interact with diagnostic assistance systems is necessary to increase throughput and reduce costs while maintaining accuracy. More importantly, to improve the problem-solving ability, operationalizable ML algorithms must be compatible with the characteristics of the data and the requirements of users [186]. Section 2.2 reviews the latest investigations relevant to representative supervised ML algorithms and further discusses their impact on clinical medical disease diagnosis and healthcare applications.

2.2 Supervised ML Techniques

In artificially intelligent systems, the purpose of ML-based diagnostic tools is to continuously approximate the ability of high-precision prognosis that can be accomplished by experienced physicians [7, 64]. By investigating the strengths and weaknesses of each ML algorithm and combining them with data characteristics of RD questionnaires and the scientific research purpose, supervised learning algorithms that are dedicated to predicting qualitative responses through pattern recognition have gained attention. Supervised learning algorithms guide and train the learning process in a data set and iteratively make predictions regarding the training data, which are corrected by the supervisor [193]. The learning process ends when the algorithm achieves an acceptable level of performance [181]. The predictions of supervised learning involve selecting feature combinations among subgroups that best describe new instances, discovering new relationships, and estimating unknown parameters [198]. Since the patients' data records collected from RD questionnaires are associated with a diagnosis category (discrete class labels), the mission considered by the supervised learning is a classification issue that designates the corresponding class label to newly observed samples by learning relevant parameters and labels from sufficient and valid reference records [44, 173].

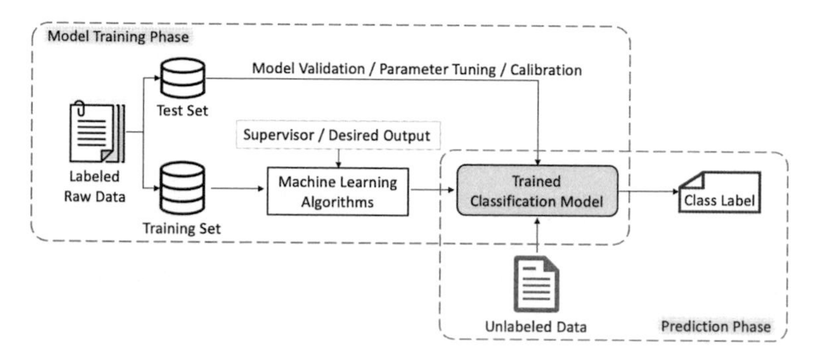

Figure 2.2: Supervised Classification Workflow: From Training Machine Learning Models to Making Predictions on New Data

The supervised classification process can be divided into the model training phase (inference phase) and prediction phase (decision phase). As shown in Figure 2.2, the labeled raw data are divided into a training set for model training and a test set for evaluating the model's performance. During the model training stage, the supervised ML algorithm analyzes the training data with assigned labels to generate a discrete prediction inference function (classification model) [250]. The test data can be used for model validation, parameter tuning, and calibration to achieve the required model performance. The prediction phase uses the classification model gained from the inference stage to compute the posterior probabilities and predict the possible category of previously unseen instances. By determining the answer patterns and predicting output results based on possible input values, the supervised classification methods meet the requirement of extracting features from existing data sources and reasonably apply to non-observed situations.

From the perspective of learning style and decision-making process, the supervised classification methods can be subdivided into discriminative and generative approaches. The discriminative classification approach utilizes the category identification information of the training data to model the decision boundaries between different classes. It directly learns the conditional distribution to identify the optimal classification surface and computes the posterior probability to distinguish the class label of an instance. Discriminative models only learn the parameters that can maximize the conditional probability to simplify the learning process but cannot reflect the overall characteristics of the training data [190]. Typical discriminative classification models are logistic regression (LR), support vector machines (SVM), k-nearest neighbors (KNN), linear discriminative analysis(LDA), and traditional neural networks [102]. On the contrary, a generative modeling approach explicitly simulates the actual distribution of the data points and the characteristics of each category to describe the phenomenon or underlying patterns in data.

Generative classifiers generate joint probability distribution and prior probability distribution from the training data and use Bayes rules to evaluate the conditional probability distribution in order to make predictions [25, 102]. The generative model can be a powerful feature extraction tool to select indicative characteristics for promoting classification accuracy. The representative generative models are naïve Bayes (NB), Gaussian mixture models (GMM), hidden Markov models (HMM), and generative adversarial networks (GANs).

In general, the discriminative classifier only provides a prediction model for the target variable, while the generative method generates a complete probabilistic model for all variables. Discriminant models are more widely used in many applications because they can provide good predictive performance and are more robust to outliers, whereas generative models can converge to the true distribution pattern of the data if sufficient training samples are provided [25]. Jain et al. [119] and Fernández-Delgado et al. [81] summarized some well-known ML algorithms and the process of selecting appropriate classifiers. Considering the research purpose and the characteristics of experimental data, the following lists some eligible ML algorithms for classification with a brief introduction of their strengths and weaknesses.

- **Logistic Regression**
  Logistic regression (LR) is a discriminative model accompanied by several model regularization methods. LR maps the prediction value between 0 to 1 through a sigmoid function so that the predicted value can be viewed as the probability of a specific category. It has no assumptions on the distribution of the features, but requested classes are linearly separable. As a widely used classifier, LR is relatively robust, and the output of LR directly yields a probability for each category [207]. When the focus is on the relationship between a single dependent variable and several independent variables, LR estimates the conditional expectation of the dependent variable when any one of the independent variables is changing.

  LR has enhanced features and functionality for binomial and polynomial classifications. It has the advantages of being easy to understand and implement, fast computing, low calculation cost, and the ability to readily observe the sample probability score. However, when it faces large feature space, the performance of LR is not satisfying and has poor handling of outliers while having a high bias.

- **Support Vector Machines**
  A support vector machine (SVM) is a discriminative classifier defined by a separate hyperplane, which can correctly divide the training data set at the maximum geometric interval in the feature space [248]. The principle of SVMs is mapping the points from low-dimensional space to high-

dimensional space to make them linearly separable. An important property of SVM after training the model is that most training samples do not need to be retained. The final classification model is only relevant to support vectors used to determine the optimal hyperplane, which distinguishes different classes by finding maximum margin [57]. The obtained hyperplane is used to classify newly observed data, where the category of the new data depends on which side of the decision hyperplane that the data point falls on. SVM is a binary classification model that can be used for classification, regression, and outlier detection. By applying appropriate kernel functions (e.g., linear, non-linear, radial basis function, polynomial, and sigmoid) to transform input data into the required form, the SVM algorithm can solve both linear and non-linear problems [121].

There are several advantages of using the SVM classifier. First, it is more effective in high dimensional spaces and can handle non-linear problems very well. Second, it is suitable for ML under small samples. Third, it has a strong generalization ability in practice and has a low risk of overfitting [11]. On the contrary, SVMs have weak interpretability on the high-dimensional mapping of kernel functions and high memory consumption but low efficiency when observing large samples. What's more, there is no universal solution to non-linear classification problems, and, sometimes, it is difficult to find a suitable kernel function. Since the SVM classifier operates by placing data points on one side of the hyperplane, there is no probability explanation [206].

- **Linear Discriminant Analysis**
  Linear discriminant analysis (LDA) is a linear projection transformation technique mainly used for dimensionality reduction. Meanwhile, it also supports both binary and multi-class classification. In a given training data set, LDA tries to project examples onto a straight line to make the projection points of similar cases as close as possible, while the projection points of different categories are as far away as possible [180, 280]. The LDA algorithm uses Bayes' theorem to estimate the probabilities of the new input data that belong to each class. The category that obtains the highest probability is the model's predicted category [121]. The goal is to project the data set onto a lower dimensional space with good class separability and find a new feature subspace that retains most of the category distinction information to avoid overfitting and reducing computational costs [116, 145].

  Flowing from Fisher's linear discriminant, LDA is a well understood and effective classification method that helps visualize high-dimensional data or simplify data structures [45]. It maximizes the between-class scatter while minimizing the within-class scatter without assumptions on data [14]. As a supervised dimensionality reduction method that uses labels and categories to measure differences, LDA has a relatively clear purpose — as opposed to principal components analysis (PCA) — and can distinctly reflect the differences between classes. LDA works efficiently when different categories have equal covariance, and the population fulfills multivariate normal distribution. However, LDA is not suitable for dimensionality reduction while preserving complex structures of non-Gaussian distribution samples [279]. If the dimension is much higher than the number of samples in the data matrix, LDA cannot find the space of lower dimension, and cannot distinguish non-linearly separable categories [253].

- **K-Nearest Neighbors**
  The k-nearest neighbors (KNN) algorithm is a supervised ML algorithm that can solve either classification problems or regression problems [173, 179, 271]. The idea of the KNN algorithm is to measure the Euclidean or Manhattan distance between different feature values. For a new input instance, KNN locates k data records in the training set that are closest to the designated test instance, where k specifies the number of neighbors to be used [250]. If the majority of these k instances belong to a particular category, then the test sample is classified into the same category [29, 131]. Different k values might lead to significantly different classification results, and the optimum k value can be determined by applying the cross-validation method. Due to the measurement unit and range differences of various features, it is necessary to normalize the feature variables so that each feature is considered equally important in the classification [121].

  As a lazy learning algorithm, KNN neither makes assumptions on data nor requires model training, which makes the KNN algorithm much faster than the algorithms that have a training period. In addition, new data can be added into the data set unremittingly without affecting the accuracy of the algorithm. But KNN is not interpretable, and it is sensitive to noisy data, missing values, and outliers [255]. Since KNN calculates the distance between the new case and each sample in the data set, the calculation complexity is proportional to the size of the data set and the number of feature variables, while the computation speed is inversely proportional. The accuracy of the KNN classification is affected when there are overlaps between categories, and significant differences between different classes' sample sizes will cause prediction deviation. Therefore, KNN is more suitable for small amounts of data with low accuracy requirements [71].

- **Decision Trees**
  Decision trees (DTs) are non-parametric supervised learning algorithms that predicts the classification or regression results of the target instance by learning a series of hierarchical decision rules inferred from characteristics of the data set [121]. DTs follow the divide-and-conquer strategy, the decisions made at internal nodes are the branching criteria, and each leaf node is assigned to one category or its probability [198]. DTs are prone to overfitting during the process of growth, which reduces the generalization ability. The primary means to deal with overfitting is pruning. Pre-pruning evaluates the situation before considering whether to branch during the construction process, and post-pruning evaluates the performance of DTs and the necessity of branching with a bottom-up approach after constructing a complete decision tree [27, 135, 182].

  DTs are easy to interpret and visualize. As a flowchart-like tree structure, DTs can fast classify new records, and the output results are easy to understand. The DT algorithm can also handle high-dimensional data with sufficient accuracy and easily capture non-linear patterns. Furthermore, it has low data pre-processing requirements — for example, there is no need for data generalization, such as removing redundant or blank attributes [182]. But the shortcoming of the DT algorithm is that it is sensitive to noisy data, neglects the correlation between attributes in the data set, and is very time-consuming to construct. The small variation in the training data can result in a different decision tree. In an imbalanced data set (the sample size of different categories is inconsistent), the information gain is biased towards the features with more numerical values, so it is essential to process the data set before creating a DTs model [68]. Apart from that, the DTs model does not support online learning. If new samples are added to the training data set, the DTs model must be completely reconstructed. Another disadvantage of DTs is that they are easy to overfit, but it is also the case with integration methods, such as random forests.

- **Random Forest**
  Random forest (RF) is a type of "bagging" (bootstrap aggregation) method in the ensemble learning technique that aggregates the results of many randomly constructed decision trees [121]. RF uses a random sample from the training data set to independently train each decision tree and randomly select features as splitting nodes to build the decision tree. Each decision tree in the forest will classify the new input sample separately and get its classification result. The final prediction of the unobserved data is made by the majority voting on several individual trees [212]. Introducing randomness when constructing decision trees and averaging the results can reduce the affection of random fluctuations in the training data set on the final outcome. Meanwhile, it increases the capability of pattern generalization [32].

  An RF can process high-dimensional data without feature selection because feature subsets are randomly chosen. In the RF, decision trees are independent of each other and trained in parallel, which leads to high training speed and the ability to detect mutual influences and correlations between features [197]. The RF algorithm has a strong anti-interference ability. It can balance errors in an unevenly distributed data set and maintain the accuracy of decision-making even with a significant number of missing features. Practices have proven that RFs are more robust and less prone to overfitting than a single decision tree [160], and the error rate is further reduced. But RFs may have low performance with small or low-dimensional data sets, and there might be numerous similar decision trees in the forest, causing the actual results to be concealed. In addition, although the data execution process of RF is faster than boosting, it is much slower than a single decision tree.

- **Naïve Bayes**
  Bayes' theorem finds the probability of an event by the given probability of another event that has already occurred. Based on Bayes' theorem, naïve Bayes (NB) classification algorithms use probability and statistics knowledge to calculate the posterior probability through the prior probability. The class corresponding to the maximum posterior probability is selected as the final classification result [125]. NB algorithms assume that features used for classification are all independent, and each feature is given the same importance (weight) to contribute to the result equally; there are no inconsequential fea-

tures. Popular NB classifiers are Gaussian naïve Bayes, multinomial naïve Bayes, and Bernoulli naïve Bayes, the main difference is they make different assumptions about the distribution of conditional probability [121, 222].

NB algorithms are relatively simple and perform well in practice. The NB model is highly beneficial for large data sets because it is easy to construct and only requires a small quantity of training data to estimate the indispensable parameters for classification. Under the conditional independence assumption, the learning and classification process of NB can be speedy compared to sophisticated methods, and it converges faster than discriminative models, such as LR [135]. But the most significant disadvantage of NB is the assumption of independent predictors, which limits the classifier's performance — in most real-life situations, there are dependencies between predictors [190].

- **Artificial Neural Networks**
  Artificial neural networks (ANNs), also known as multi-layer perceptrons, abstract and imitate fundamental characteristics of human learning and cognition process to approach the level of artificial intelligence. The essence of the ANN method is to simulate the relational function between features and target variables. ANNs consist of an input layer, an output layer, and a hidden layer. The input layer contains nodes representing features; the output layer covers nodes that interpret predicted classes' probabilities, and the sum is 1. Between the input and output layers, the hidden layer consists of neurons that can be viewed as computation and storage units. Neurons execute the calculation function for the input value, temporarily store the results, and transfer them to the next layer [121]. ANNs accomplishes the processing and transmission of information through cooperation between neurons. The representation of ANNs is a directed graph. The ANN model is parameterized by assigning weight to each arrow, and the purpose of model training is to adjust the optimal weight to reach the most effective prediction [104].

  The attractiveness of ANNs lies in their magnificent information processing characteristics. The ANN algorithm has strong robustness and fault tolerance because all quantitative and qualitative information is equally distributed and stored in neurons. It can sufficiently approximate complicated non-linear relationships. The highly parallel distributed processing enables performing a considerable amount of operations rapidly. The self-learning and generalization capabilities bring the model convenience to adapt to unknown or uncertain systems. However, ANNs are challenging to explain, and constructing a neural network requires preparing large volumes of data in advance. Model training can be very time-consuming because the selection of parameters (e.g., network topology, weights, and thresholds) relies mainly on experiences [19] [120].

Different ML methods have different strengths and weaknesses. Their performance is primarily dictated by the characteristics of the data to be classified, and no single classifier can perform well for all given problems. Osisanwo et al. [198] compared seven different classification algorithms using the diabetes data set and determined the relatively effective algorithm based on the characteristics of data. Their results show that SVM is considered to be the most accurate algorithm, followed by NB and RF. Meanwhile, a vital research conclusion is proposed: ML algorithms require accuracy, precision, and minimal errors to have exceptional performance regarding supervised prediction.

2.3 Classification Model Evaluation and Selection

The ultimate goal of ML is to efficaciously generalize universal rules and potential knowledge that can be applied to predict unseen cases. The generalization performance of a trained model reflects the adaptability of making appropriate predictions for new samples that have similar characteristics to the training data [241, 278]. By utilizing selected algorithms and collected data, the ML task establishes a realistic hypothesis model to learn the general concepts of a phenomenon and provide a solution that depends on the characteristics of the input data from the problem domain [7]. Therefore, the ML model's generalization ability is influenced by the learning capacity of the algorithm, the sufficiency and accuracy of the data, and the difficulty level of the learning task [117].

However, the collected training data may contain irrelevant noise or disturbances because they were recorded from partial experience fragments. Moreover, the data distribution or characteristics under the research topic may drift over time. The features of the training samples cannot fully represent the general properties of all potential cases in the actual sample space. Only a partial view of the research subject is inadequate to train a precise diagnostic model to comprehensively cover all possible scenarios, which will lead to a deviation (error) between the learner's prediction output and the sample's actual situation [287]. Four types of prediction errors could occur during the model training phase: mislabeling, representation, learner, and boundary errors. Nevertheless, different from the prediction errors from the training set, the generalization error is obtained from the model's prediction on objects that have not participated in the training set [176].

The generalization error of a model can be decomposed into bias, variance, and noise (irreducible error) [215]. Noise is the inevitable interference information contained within the collected data in practical applications [95]. The occurrences of noise or random fluctuations could impact the classification model's accuracy and prediction results [129]. In an ML task, insufficient training in the early stages leads to poor fitting ability of the learner. If a learning model is too simple to capture the underlying correlations between the independent and dependent variables, bias will dominate the generalization error rate and lead to underfitting. As the learning model continuously trains to increase the complexity, the model's fitting ability gradually enhances, and the variance stepwise becomes the leading factor of the generalization error rate.

A complex model can perform well with training data and has robust fitting capabilities. However, if the model is overtrained, it may regard some non-global and unique characteristics of the training samples as the general properties of the data set. Excessive complexity of a learning model will reduce the generalization ability and bring about overfitting. If irrelevant noise or random fluctuations in the training data are learned as concepts by the model, these irrelevant information will negatively impacting the model's generalization performance and predicting ability on new instances [21, 265].

For a supervised learning task, because noise is an irreducible error, the cause of poor generalization performance is either overfitting or underfitting. Thus the primary focus of model optimization is to find an appropriate balance between bias and variance [117]. However, bias and variance are subtended, known as the "bias-variance dilemma". Underfitting models usually have high bias and low variance, and overfitting models have low bias and high variance. To achieve adequate generalization performance, the bias must be small, which means that the data can be adequately fitted. And the variance must be small, which means that the impact of data disturbance is small.

The "bias – variance trade-off" implies that a model should reconcile underfitting and overfitting to ensure that the model is rich enough to express underlying structures of learned data and achieve the required performance level but simple enough to avoid fitting unnecessary characteristics [21]. The optimum balance point is the position before the error on the test set starts to increase where the model still has good performance on both the training set and test set. In order to obtain a correct ML model for effectively conducting statistical analysis, prediction, and decision-making, it is essential to analytically evaluate the capability and accuracy of the designed model in predicting reality. Based on the data set size and data distribution, this section introduces three commonly used data partitioning and model evaluation mechanisms, as well as essential evaluation metrics for measuring the generalization ability of a classification model.

2.3.1 Model Evaluation Mechanisms

Due to the existence of overfitting, the error in the training data is inadequate to be used as a measurement of model accuracy, and future samples that are supposed to calculate the generalization error are unavailable. In practical applications, the primary strategy of model evaluation is to partition a portion of training samples into validation samples. This means that the entire data set is divided into three parts: a training set, a validation set, and a test set. The training set is used to train the classification model, and the validation set is used to measure how well the model is trained. Meanwhile, some learner's hyperparameters can be selected and adjusted through the evaluation indicators on the validation set.

Different learning models can be compared by the error rate on the validation set, and the model with the least error is chosen as the optimal learning model. Then, the training and validation sets are merged to retrain the selected model. Finally, the test set is used to simulate the data that the classification model encounters in real-world applications [143, 287]. Particularly, suppose that there are no hyperparameters to tune or only experimental data and no requirements for product deployment. In that case, the experimental data only need to be split into training and test sets to evaluate a concrete model's performance. Since the test set and the training set are independent of each other, the prediction error generated by test samples can approximately represent the model's generalization error [215].

Model evaluation provides intuitive results to illustrate the performance of the learner. The smaller the test error, the stronger the model's generalization ability [136]. According to different data segmentation methods, the following paragraphs enumerate three commonly used mechanisms for assessing a model's generalization ability. The comparison of these three methods is shown in Figure 2.3.

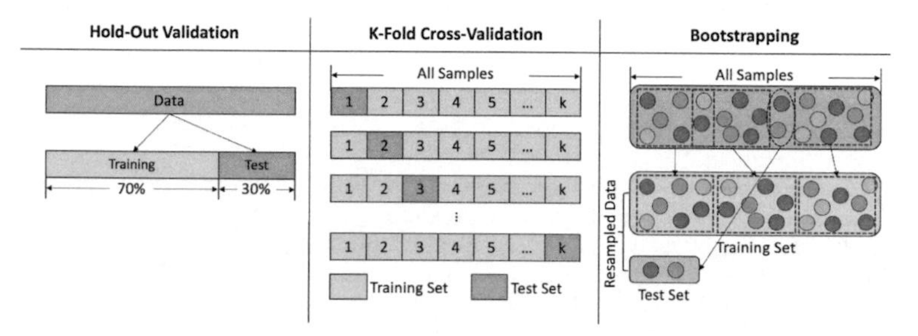

Figure 2.3: Principle Techniques for Evaluating ML Models: Hold-Out Validation, K-Fold Cross-Validation, and Bootstrapping

- **The hold-out method** randomly divides the raw data set into two mutually exclusive subsets. Different division schemes lead to different components in the training or test set, and the sampling differences will affect the corresponding model training and validation results. A small test set will bring about a significant variance in the evaluation results, which reduces the fidelity and persuasion of the obtained learning model. In contrast, a small training set will lead to a high bias on sample estimation, which leads to accuracy reduction. Usually, 70% of the raw data are designated as the training set and the remaining 30% as the test set. In classification problems, in order to avoid additional bias introduced by the data partitioning process, stratified sampling is applied to ensure that the training and test sets maintain the same proportion of categories as the original data set. Both subsets are independent but identically distributed. Even though the hold-out method has the advantage of lower computational cost, the performance of the model is subject to higher variance when the total sample size is small, so the generalization error of the learned model is less reliable [206]. In practical applications, the hold-out method is usually applied multiple times with random division to obtain a more precise evaluation of a classifier's accuracy and error rate by taking the average of repeated model evaluation results [287].

- **K-fold cross-validation** randomly but equally partitions the raw data set into k mutually exclusive subsets. By iteratively using each subset as a test set, each sample has an opportunity to become a test sample, while the remaining $k-1$ subsets are used for training the model [276]. The performance of the learning model is determined by averaging the prediction error rates of k test sets. The k-fold cross-validation allows all data to participate in the training process. For learning tasks with limited sample size, the training data distribution may deviate from the actual distribution, and using all data to train the model will cause overfitting. On the contrary, k-fold cross-validation effectively reduces bias and alleviates overfitting through k iterations [284].

In the k-fold cross-validation, the stability and fidelity of the evaluation results are highly dependent on the value of k [284]. The range of k is greater than 1 and less than or equal to the sample size N. There are two commonly used k values: $k = 10$, namely 10-fold cross-validation; and $k = N$, also known as leave-one-out cross-validation (LOOCV). In LOOCV, the iteration is repeats N times that every sample can be used as test data, while the remaining $N - 1$ samples are used for training the prediction model [45, 276]. The advantage of the LOOCV is that it is not affected by the random sample division method. Furthermore, the evaluation result is considered to be more accurate because the distribution of training data is asymptotically equivalent to the original data set, with only one sample difference [286].

However, the shortcomings are also obvious. The calculation speed will be greatly reduced by training N models, especially on large data sets, and the computational complexity and calculation cost are considerably high, so LOOCV is rarely used in general practical applications [127]. For a stable learning model, the results of the LOOCV are more uniform and reliable compared with k-fold cross-validation. However, if the model has low stability, a higher k value can obtain better estimation results, though from an experimental point of view, increasing the value of k does not necessarily reduce the variance [144]. The value of k still needs to have a case-by-case analysis.

- **Bootstrapping** is a repetitive random sampling approach with replacement. The obtained bootstrapped data set is of equal size to the original data set, but some samples are repeatedly selected and appear multiple times in the training set, while some instances might be omitted completely [73]. The data points that have not been captured from bootstrap sampling are assumed to be the test set. The bootstrapping method can construct multiple different training sets from the original data set. It is useful for a relatively small data set that is difficult to partition into training or test sets, and it brings significant benefits for methods such as ensemble learning. However, the generated data set changes the distribution of the original data set, which introduces estimation bias [73]. Therefore, if the amount of raw data is sufficient, the hold-out method and k-fold cross-validation are more frequently used [17].

2.3.2 Evaluation Metrics for Supervised Learning Models

The performance metrics are the evaluation criteria for measuring the generalization ability of a learning model. The selection of metrics depends on the different types of ML tasks and directly influences the method of measuring a model's performance in predicting the outcomes. For classification model evaluation, the confusion matrix is often used to derive assessment metrics, such as accuracy, sensitivity, specificity, and precision. The logarithmic loss or cross-entropy loss (log-loss) and either the ROC or the area under ROC curve (AUC) are also frequently used for appraising the classification results from different perspectives [242]. For regression model validation, the mean absolute error (MAE), mean squared error (MSE), root mean squared error (RMSE), R^2 are regularly used indicators. In addition, the precision-recall curve and normalized discounted cumulative gain are used for measuring the query or ranking quality in web search engine algorithms.

Consider a binary classification problem that classifies instances into a positive or negative class. There are four situations according to the combination of an instance's actual category and the estimator's predicted category. If an example is actually positive and also predicted as positive, it is a true positive (TP). If the instance is in the negative class but predicted as positive, it is a false positive (FP). Correspondingly, a true negative (TN) is a negative instance that is also predicted as negative, while a false negative (FN) is a positive instance but predicted to be negative [242].

Table 2.1: Assessing Model Performance with Confusion Matrix and Evaluation Indicators [242]

		Predicted Class		
		Positive	Negative	
Actual Class	Positive	True Positive (TP)	False Negative (FN)	$TPR = Sensitivity = Recall = \frac{TP}{TP+FN}$
	Negative	False Positive (FP)	True Negative (TN)	$FPR = \frac{FP}{FP+TN}$; $Sepcificity = \frac{TN}{FP+TN}$
		$Precision =$ $PPV = \frac{TP}{TP+FP}$	$NPV = \frac{TN}{TN+FN}$	$Accuracy = \frac{TP+TN}{TP+TN+FP+FN}$

These four situations form a confusion matrix, as shown in Table 2.1. *Accuracy* indicates the correctness of the ML algorithm when classifying TP and TN samples. Accuracy is useful for samples that are uniformly distributed, but is not a good choice for unbalanced data. Accordingly, the *Error Rate* is the proportion of misclassified samples to the total amount of samples, which is equal to $1 - Accuracy$. *Precision*, also commonly called *Positive Predictive Value (PPV)*, is the proportion of correctly classified positive examples (TP) to all positive observations that the model predicted. The *Negative Predictive Value (NPV)* is the probability of subjects that are classified with negative results that are truly negative.

Sensitivity, also known as the *True Positive Rate (TPR)* or *Recall*, depicts the proportion of positive instances identified by the classifier to all TP instances in the data set. The greater the sensitivity value, the more accurate the classification algorithm is at classifying the true positive samples. *Specificity* indicates the accuracy of the algorithm at correctly identifying TN samples in all negative samples. In contrast, the *False Positive Rate (FPR)* delineates the percentage of negative instances misdiagnosed by the predictor as a positive class in all negative instances, which is equal to $1-$ *Specificity*. The greater the specificity, the more precise the algorithm is at classifying TN samples [42, 112, 242]. In medical diagnostic research, sensitivity and specificity are two important aspects of the prediction model. Sensitivity demonstrates the ability of the diagnostic system to correctly identify individuals with the disease, while specificity reflects the system's capacity to accurately exclude individuals who do not have the disease [204].

Under ideal conditions, both sensitivity and specificity are expected to be high, which means that the TPR value should be close to 1, and FPR should be close to 0. In practical application, the objective is to find an equilibrium point between sensitivity and specificity. This process can be represented by the ROC curve, which is an aggregative indicator that graphically reveals the relationship between sensitivity and specificity [77]. If the output of a binary classifier is the probability value of predicting positive samples, the setting of different thresholds will lead to different confusion matrices. The ROC curve plots TPR on the y-axis and FPR on the x-axis. To generate the curve, the classifier's predicted probability results sort all samples, and each sample's predicted probability is used as a threshold to calculate the corresponding FPR and TPR. The ROC curve assesses the relative changes between TPR and FPR across a range of threshold values. The upper left corner of the graph, where $FPR = 0$ and $TPR = 1$, represents the critical value where the model classifies all samples correctly. A ROC curve that is closer to the upper left corner indicates higher sensitivity and specificity, and a better performance of the classification model. Conversely, a curve closer to the diagonal suggests more random classification [100].

ROC curves can intuitively observe the influence of different thresholds on the learner's generalization ability. In bioinformatics, since the ROC curve does not have a fixed threshold and allows intermediate states, it is helpful for researchers to weight the impact of misdiagnosis in combination with their professional knowledge, thus choosing an optimal threshold as the diagnostic reference value [148]. Furthermore, by drawing the ROC curve of different learning models to the same coordinate, the performance quality of different learning models can be visually distinguished. The closer a learner's ROC curve is to the upper left corner, the higher the classification precision is [148]. Meanwhile, the AUC calculates the two-dimensional area underneath the ROC curve to provide a comprehensive performance measure across all possible classification thresholds. The AUC represents the probability of a classifier ranks a random positive sample higher than a random negative sample. The range of the AUC is from 0 to 1, that is to say, a model that has 100% wrong predictions will have an AUC value of 0.0, and the model with 100% correct predictions will have an AUC value of 1.0 [39]. The higher the AUC value, the more precisely a model can distinguish between positive and negative subjects. Since the AUC metric has characteristics that are independent of the selected threshold or the scale of probabilities, it can make a realistic evaluation of the classifier — even in the case of an unbalanced data set [100].

Similarly, a PR curve is a plot of the precision (on the ordinate axis) and the recall (on the abscissa axis) for different threshold values. The top right corner of the graph is the ideal space for getting the optimum condition [213]. Due to the absence of a TN value in the precision and recall equation, the PR curve is more informative than the ROC curve in evaluating binary classifiers on unevenly distributed data sets [229]. Especially when detecting the positive class as essential, the negative class dominates. For example, when diagnosing rare disorders, among all patients with similar symptoms, only a few cases were diagnosed with an RD (TP). Therefore, the collected data are often highly imbalanced. However, medical specialists want to ensure that patients who the system predicts indeed suffer from an RD (precision), and they do not want to miss any patient with an undiagnosed RD (recall). Under circumstances in which the majority of observations are TN, the ROC curve may exhibit an overly optimistic perspective of a classifier's performance due to the consideration of TN in the FPR equation. While PR curve only evaluates the fraction of TP among positive predictions and avoids TN values [61].

Another summative composite score to measure the accuracy of a classification model is the *F1 score*, which calculates the harmonic mean of precision and recall. The higher the F1 score, the better the model's performance [242]. Additionally, *log-loss* quantifies the accuracy of a classifier by penalizing misclassifications. The classifier must assign a probability value to each predicted category instead of declaring the most likely class to calculate the log-loss value. The range of log-loss is $[0, \infty)$, minimizing log-loss is equivalent to maximizing the classifier's accuracy. Before applying a trained classification model for predicting future data, different performance evaluation metrics provide a quantitative and intuitive explanation of the generalization ability of the model [204]. In addition, by setting a particular evaluation index as the training target of a learning model, the pre-defined value of the indicator can be used as a benchmark for continuous optimization of the model until the desired value is achieved. It is worth mentioning that the evaluation metrics are different from the loss function (or cost function). The loss function estimates the degree of discrepancy between the model's predicted output and the sample's actual value during the training procedure. In contrast, evaluation metrics judge the model's performance after the learning process. However, both the loss function and evaluation metrics have same objective of finding the optimal model that fits the learning task and training set. Decreasing the loss error will lead to the improvement of evaluation metric values [287].

2.3.3 Statistical Hypothesis Testing for Algorithm Comparison

The experimental model evaluation mechanisms, such as k-fold cross-validation, train and test k models to calculate the mean performance metrics to obtain a comprehensive assessment of a trained classification model. However, the measurement of different learners' generalization abilities cannot be directly obtained by comparing the performance metrics because, first of all, the data distribution in the natural environment is unpredictable, while the model's performance metrics are obtained from the test data set. The generalization error acquired on the test set is an approximate conjecture of the real-world situation but is not equivalent to reality. The performance metric comparison results can be misleading, as it is difficult to know whether the difference between the mean performance scores is statistically significant or caused by noise or coincidence. Second, the model's validation results on the test set are highly correlated to the randomness in sampling, observation subject order, and the test data set size. The model performance indicators inferred from the random samples are probabilistic, not absolute positives or negatives. Third, the stochastic of some ML algorithms, such as random forests, makes the model produce discrepant results and even execute the validation process multiple times on the same test set with the same parameters [31, 215, 287].

Statistical hypothesis testing provides an essential foundation for comparing the generalization ability of different ML algorithms on a target classification task [66, 97]. Especially when the prediction accuracies of multiple learning models are similar and difficult to choose, the statistical hypothesis testing verifies whether the observed performance differences are statistically significant to assist the scientific comparison process [37, 97]. By claiming two mutually exclusive hypothetical propositions (*Null Hypothesis* $\boldsymbol{H_0}$ and *Alternate Hypothesis* $\boldsymbol{H_1}$) about the expected mean generalization error differences of both learners across all study populations, the hypothesis testing attempts to provide statistical evidence for determining the plausibility of the conjecture (null hypothesis) through test data [234].

Inferential statistics describe that if the sample size is sufficient, the population's characteristics will be revealed in the test set [164]. Assuming the samples are independently distributed, then the characteristics or distribution of the test samples should be consistent with the population parameters (or distribution) in the sample space. Therefore, the distribution of the generalization error can be estimated by the test error. The idea of comparing the generalization performance of two different learners is that if two ML algorithms perform the same, then both classification models trained by the same data set should have the same error rate, and, thus, the difference between two proportions should have a mean value equal to zero. According to the hypothesis testing results, it can be inferred that the credibility of learner A's generalization performance is statistically better than learner B if learner A outperforms B on the test set [215].

Whether to reject the null hypothesis is based on the evidence provided by the randomly selected sample from the population of interest, the test statistic, and the decision rule. The formulated decision rule defines rejection regions that tell under what circumstances to reject the null hypothesis, depending on the research hypothesis and the predetermined level of significance. A hypothesis test's *significance level* (α) is a predetermined probability threshold that limits the maximum probability of rejecting the true null hypothesis. The frequently utilized values for significance level are $\mathbf{0.05}$, $\mathbf{0.01}$, and $\mathbf{0.001}$. For example, $\alpha = \mathbf{0.05}$ indicates that a chance of $\mathbf{0.05}$ to improperly deny $\boldsymbol{H_0}$ is acceptable [24]. The smaller the selected α, the more robust the interpretation of $\boldsymbol{H_0}$ is [252].

The *test statistic* and p-value are used to interpret whether the difference is caused by statistical chance or essential inconsistency between the hypothesis and the actual population. A test statistic is a number that summarizes the sample information to measure the degree of agreement between observed data and the expected distribution under the null hypothesis. Different hypothesis tests use different test statistics depending on the probability model assumed in the null hypothesis [130]. The commonly used test statistics include the t-value in a t-test, z-value in a z-test, χ^2-value in a chi-square test, and F-value in an analysis of variance (ANOVA) test [130]. The p-value is a probabilistic value calculated by test statistics. By comparing the p-value to a *significance level* (α), the p-value quantitatively measures which assumption is better supported by the sample data, thus deciding whether to reject H_0. Assuming H_0 is true, p-value $\leq \alpha$ means that the selected sample leads to a low probability event. Thus H_0 will be rejected because the test statistic shows remarkable abnormal evidence. The H_0 will not be rejected if α is smaller than p-value, which means that it is not a significant result. The smaller the p-value relative to α, the more substantial the evidence to reject H_0 is [15].

In practical experiments on small data sets, the k-fold cross-validation t-tests are often advocated to operate the training or testing process multiple times and gather a set of performance measures to reduce the uncertainty of a model [97, 261]. For example, to compare the performance of RF and LR models on the same learning task, the data are split into k folds to train and test k RF models and k LR models. Both learning models yield k corresponding performance metric values, such as error rate, for the statistical testing. By calculating the difference value between each pair of performance indicators and finding the mean and variance of the k pairs of differences, a t-distribution with $k - 1$ degrees of freedom can be constructed. Whether the observed difference is statistically significant can be inspected by calculating the mathematical characteristics of the distribution. The k-fold cross-validation avoids the problem of overlapping samples on the test set since there is no intersection of test samples. However, the training sets of different rounds will overlap due to continuous resampling of the same data set. For instance, in 10-fold cross-validation, each round of model training will share 80% of the training data with other rounds. This means that the obtained performance metric values are not independent, which violates the assumption of statistical hypothesis testing and leads to an overestimation of the probability that the null hypothesis is tenable.

Thomas G. Dietterich [66] reviewed five commonly used significance tests to compare different learning algorithms' generalization performance by measuring their type I error and type II error on artificial and real-world data. A type I error (false positive) occurs if a true null hypothesis is incorrectly rejected, resulting in unnecessary changes or interventions that waste time and resources. The type II error (false negative) is when there is a difference, but the null hypothesis is not rejected. The likelihood and impact of type I and type II errors must be weighed based on different research emphases because formulating stricter criteria in rejecting the null hypothesis can reduce type II errors, but the adjustment will increase the chance of type I errors.

Experimental results suggested that the five replications of the 2-fold cross-validation (5×2 CV) t-test is the most powerful technique, followed by McNemar's test if the learning algorithm can only run once. Gitte Vanwinckelen and Hendrik Blockeel [261] mentioned that the bias in the estimated mean accuracy due to the k models being trained from a subset of the collected data could be minimized by performing LOOCV. Furthermore, the variance caused by differences in random sample partitioning can be reduced by repeating the k-fold cross-validation n times to compute the average of multiple performance metrics. Remco R. Bouckaert and Eibe Frank [37] studied the consistency and replicability of three popular significance tests. The experimental results indicate that **10**-times **10**-fold cross-validation has appropriate type I errors and low type II errors, with sufficient reproducibility to reproduce the published results.

2.4 Ensemble Methods for Classifiers

The "no free lunch" theorem states that commonly used classification algorithms have inevitable errors, different classifiers have inequable peculiarities, and no classifier can be universally applicable to all types of ML issues [230, 274, 275]. The classifier ensemble approach, which constructs a group of base learning algorithms that are individually trained, can improve the system's overall prediction performance and optimize the ML results by combining several classifiers' estimation results [67]. Ho et al. [110] stated that because of the complementary recognition methods and flexibility in dynamic adaptation, a suitable integration of various classifiers can achieve more significant generalization ability or robustness and accuracy than any constituent model.

According to the type of base learners that constitute the ensemble algorithm, a homogeneous ensemble method contains only one type of classifier, such as decision trees in random forests. In contrast, the individual learners in heterogeneous ensembles are generated by different types of learning algorithms [86, 194]. When selecting learning algorithms for the heterogeneous ensemble method, the main challenge is not to pick highly accurate classifiers, but to choose classifiers that can produce different types of induction deviation [224]. Bias–variance balancing is the crucial objective of ML. The ensemble methods can effectively use this diversity to complement each other, thus reducing variance without increasing bias [88]. Kuncheva [152] conducted an experimental study with several well-known ML algorithms to observe the suitability and potential of different combinations of classifiers. By comparing the performance of several ensemble techniques with the individual classifier, the research shows that the satisfactory performance of an ensemble system benefits from the deviation diversity of combined classifiers. The higher the diversity of misclassifications generated by individual learners, the lower their impact on the final decision. Thus, the ensemble model can achieve higher accuracy than any of its member classifiers.

The classifier ensemble methods can be divided into two categories regarding the dependence between different learners. Sequential ensemble methods generate the base model serially when there is a strong dependency between individual learners. The models in a sequential ensemble are trained in sequence and depend on the previous models. The overall performance can be improved by giving a higher weight to the misclassified instances. Parallel ensemble methods are applied to base learners that are independent of each other. Models are learned in parallel from the training set, and the error can be reduced by using the averaging method.

Table 2.2: Ensemble Learning Techniques: A Comparison of Bagging, Boosting, and Stacking [20, 83, 150, 289]

Combination Strategy	Target	Ensemble Method	Base Learner Dependency	Aggregation	Training Data
Bagging	Reduce variance	Homogeneous	**Parallel** Independent of each other	Voting method for classification Averaging method for regression	Bootstrap sampling
Boosting	Decrease bias	Homogeneous	**Sequential** Strong Correlation	Weighted majority vote for classification Weighted sum for regression	Adjust samples weight in each iteration
Stacking	Decrease both variance and bias; Improve prediction	Heterogeneous	**Parallel** Independent of each other	Meta-Learner	Training data for base learners; Basic learner's prediction results for meta-learner

There are mainly three strategies for integrating different ML techniques into one predictive model. In the homogeneous ensemble methods, bagging (the representative algorithm is random forests) focuses on reducing variance, boosting (the representative algorithm is Adaboost) focuses on decreasing bias, and stacking is a typical approach of the heterogeneous ensemble (stacked generalization) [150, 289]. Table 2.2 distinguishes the differences between bagging, boosting, and stacking in terms of targets, ensemble method, base learners, and aggregation approaches.

Bagging denotes bootstrap aggregation and uses bootstrap sampling to prepare the data sets for training base learners [40]. The method randomly selects subsets with replacements from the original training data. The extraction of the subset in each round is independent. The bagging approach is constructed by several independent learners that are trained in parallel. For aggregating the predicted results of base learners, bagging uses the voting method for classification and averaging method for regression [20]. The combined model performs better than any base learner because the variance can be reduced by averaging multiple estimates. Since the bagging method can reduce overfitting, it performs well by using strong classifiers and complex models [83, 289].

In contrast, the boosting methods are iterative algorithms that train a sequence of weak learners to generate a powerful ensemble model [83]. Boosting algorithms teach a weak learner by using instances that are assigned with equal weights. The base estimators in boosting methods are constructed sequentially, and the training set for each iteration is unchanged, but the weight of each sample in the training set is continuously adjusted according to the error rate of each estimator. Examples that are misclassified by the previous learner will obtain a higher weight before the next iteration so that the following estimator will pay special attention to high-weighted samples. The boosting process continuously uses a weak learner to remedy the deficiency of the previous learner until it reaches the number of iterations or the loss function is less than a certain threshold [20]. Each base learner is trying to reduce the bias of the combined model. The final prediction of the boosting model is integrated from each learner's predicted result through a weighted majority vote (for classification) or a weighted sum (for regression) [83, 289].

The stacking strategy is the result of further generalization of the aggregation approach in the ensemble technique. The main idea of stacking is training a model to learn the prediction results from base learners. There are two stages in the architecture of classic stacking methods. The first stage contains different classification or regression models that are learning from an entire training set. In the second stage, a meta-learner is trained by integrating base learners' outputs as input characteristics [273]. More specifically, the prediction results of base estimators are stacked together and used for computing the final prediction by the second-level estimator. The essence of the meta-learning algorithm is to simultaneously reduce the bias and variance of ML models. It can be interpreted as finding the appropriate weight for each base learner. The flexibility and indeterminacy of the stacking method make it possible to replace the averaging or voting method in bagging and boosting [150]. For instance, the averaging process can be achieved by linear regression, the voting method can be performed by KNN, and weighted voting can be obtained by LR. Stacking uses the idea of k-fold cross-validation to train the base models. The obtained out-of-fold predictions further become the foundation of the meta-model training set. The meta-learner takes advantage of a series of well-performing learning models in classification or regression tasks and makes more accurate predictions than any single model in the ensemble. Compared to independent prediction models, the stacking method is better at reducing the generalization error and increasing the depth of learning. However, the complexity of the stacking method requests additional computational cost and model interpretability.

As a compromise, the classifier fusion method is introduced to combine the decisions of all learning models.The final decision is determined by combining scores, ranks, or decision levels of multiple classifiers to obtain better prediction results [110, 151]. Extensive experimental results from Tsoumakas et al. [256] show that the fusion of appropriate predictive models is an improvement over individual learners or any simple combinations. It is noteworthy that the performance of a classifier fusion method is comparable to the stacking strategy without worrying about training the meta-learner or extra computational consumption [72, 256]. The standard approach of the classifier fusion method is to organize and train different estimators in parallel and simultaneously, and their prediction outputs are respectively taken as inputs for a combination function [165, 227]. Designing a suitable combination function for the final decision-making is the key to increasing the system's overall performance. Kuncheva [149] compared the probability of classification error of the six classifier fusion methods, with the assumption that the estimated values were independent and distributed normally or uniformly. The six combination formulas cover the minimum, maximum, average, median, majority vote, and the oracle. The study illustrates that the majority voting method has satisfactory performance on normally and uniformly distributed errors.

Among popular classifier combination schemes, the most widely-used methods originate from the voting method, such as majority voting, plurality voting, and weighted voting [277]. The voting mechanism is utilized after each learning model predicts a class type or related ranking or probability distribution. The candidate class that obtains the most votes — or the highest average ranking or probability — among many alternative classes is the final result proposed by the fusion method. The majority voting stipulates that the elected category must obtain more than 50% of the votes, and the plurality voting method selects the category that gains the highest number of votes [259]. Since different base learners contribute differently to the final classification result, the weighted voting does not treat the classification models equally. Each model is assigned a coefficient (weight), which is usually proportional to the classification accuracy of base learners [53]. Weights of voting vary between different classifiers, which can help better estimate the reliability of participating classifiers [246]. Higher weights are assigned to the learner with better performance — more accurate classifier gains higher weight [277].

De Stefano et al. [63] implemented a weighted majority voting approach that assigns weights to diverse learners to compensate for their respective weaknesses and minimizes classification errors in the classifier fusion. The experimental results show that assembling the output of each classifier according to the weighted voting combination rule enhances the classification accuracy and brings the system stronger generalization ability and universal applicability. It is worth noting that the output ranges of different classifiers are varied. For example, the output range of LR is from 0 to 1 (probability), and the result of KNN can be any number greater than 0. The output scope of other algorithms can be an unlimited range of values, including negative numbers. Therefore, different classifiers' output ranges must be normalized while integrating multiple classifiers. Moreno et al. [183] studied three weighted voting schemes for classifier fusion tasks (*re-scaled weighted vote, best-worst weighted vote, and quadratic best-worst weighted vote*). The experimental results demonstrate that the proposed ensemble methods outperform any individual classifier, and the ensemble algorithm that scales the weights to the range generated by the best and worst classifiers is manifested in most test data sets to obtain the optimum performance.

Considering the fact that an under-performing learner will hinder the overall performance of a fusion model, achieving good integration results requires screening the candidate base learners, discarding the discrepancies, and selecting appropriate classifiers. Tsoumakas et al. [256] used three multiple comparison procedures in statistics — Tukey's honest significance test, Hsu's test, and the Scott-Knott clustering algorithm — to select the best-performing group among 10 classification algorithms and applied voting and the weighted voting method to fuse the decision of the models, subsequently. The experimental analysis of 40 different data sets shows that weighted voting exhibits lower error than the voting method. By gradually increasing the number of fused algorithms from **3** to **10**, it was also found that the number of learning algorithms would not negatively influence the overall accuracy of the fusion approach. Accumulating diversified learning algorithms with high precision to the fusion framework can improve the accuracy of the model.

Classifier fusion techniques have often proven to outperform single classifiers and have been widely used in medical studies. Ozcift et al. [199] constructed a rotation forest classifier ensembles containing **30** ML algorithms to evaluate their classification performance in Parkinson's, diabetes, and heart diseases. Their study proves that the proposed classifier ensemble algorithm can be used to improve the accuracy of different ML algorithms in the design of advanced computer-aided diagnostic systems. Jaffar and M Arfan[118] presented a framework to analyze breast mammograms by using a combination of deep convolutional neural networks and SVMs. The results indicate that the proposed framework attained an accuracy of **93.35**% and a sensitivity of **93**%.

Ma et al. [166] proposed a multi-classifiers fusion method to recognize CT imaging signs of lung diseases based on classifiers' confusion matrices and related classification confidence values. The experimental results show that this method improves the recognition performance more than single classifiers. It is an efficient and promising technology that is superior to AdaBoost and bagging. Rezaei et al. [221] introduced a weighted voting classifier ensemble approach for features extraction and classification of brain tumors in MR images. A multi-objective differential evolution (MODE)-based ensemble approach was applied to obtain the classification accuracy of **92.46**%. By comparing the research achievements with the diagnosis of experienced specialists, the sensitivity and specificity indices suggest that the proposed ensemble method reflects adequate classification performance.

2.5 Probability Calibration Methods

For many applications that utilize expert systems and classification models for decision-making, obtaining a good probabilistic estimation that accurately reflects the probabilities of an instance belonging to each possible category considering all evidence is imperative to provide essential and intuitive information. Especially when physicians must determine a patient's medical diagnosis and formulate a corresponding treatment scheme with the support of the computer-aided diagnostic system, a highly accurate and well-calibrated predictive model provides risk assessments reflecting the underlying likelihood of a disease [123] and provides an understanding of the results to make final diagnoses more credible [48].

In practical problems, converting the output of a classification model into probability [0, 1] has many important implications. First, using the probability value is convenient and efficient for people to understand the output results because of the well-known fundamental concept. Second, the probability value can provide more information than categorical output (**0** or **1**), such as ordering the prediction results. The instance with a probability of **0.8** is more likely to be classified as positive than that with a probability of 0.7. Third, it helps to integrate multiple classification decisions of different models in the case of assembling various classifiers. In classifier fusion methods, the diversity of selected classifiers is an essential factor of improving the effectiveness of the prediction. However, the final decision will be significantly affected if classifiers cannot reflect the actual probability, so merely voting on classifiers' prediction results is inadequate. Converting the output range of non-probabilistic classifiers to be consistent with the actual probability distribution, the so-called probability calibration helps further revise the prediction results of a classification model to more qualified and understandable probability estimates that reflect the reality [191].

When performing statistical classification tasks, each trained classifier will generate corresponding predictive opinions, known as subjective probability, about how likely an observation is to be in a positive class, referring to its own experience and learning results. On the contrary, the objective probability is the chances of an event actually occurring, based on empirical evidence using statistical analysis, experiments, and mathematical measurements [10]. The reliability diagrams, also called a calibration plot [191, 203], is used to evaluate the calibration of the predictive model, which measures the degree of consistency between the subjective and objective probabilities. If a classifier's subjective probability is generally greater than the objective probability in predicting an event, the classifier is over-confident. Otherwise, the classifier is under-confident.

Using a synthetic data set with $100,000$ samples — from which $1,000$ instances are for model training and the remaining for testing — as an example, Figure 2.4 provides a visual assessment of the deviation between different classifiers' relative prediction scores (LR, NB, SVM, and RF) and the actual probabilities in a binary classification problem [41], where the x-axis represents the predicted probability values of a classifier in the interval from 0 to 1, and the y-axis denotes the fraction of positive records in the sample space. If a learning model is perfectly calibrated, it will yield the relationship of $y = x$ (the black dotted line appears along the diagonal), which reflects the true likelihood of the prediction. The output of a well-calibrated probabilistic classifier can interpret the classifier's confidence level.

To compute the reliability diagram of a binary classifier, the subjective probability scores of test instances to be classified as positive will be sorted and grouped into ten stacks, where each stack represents a disjoint probability interval between 0 and 1. For each subgroup of samples, two estimators need to be calculated: the mean predicted probabilities per stack (Figure 2.4 (b) shows the corresponding histogram), and the fraction of true positive records within each stack. For example, in a stack that contains 20 samples, if only one sample is a TP class, then the fraction of TP samples for that stack is $1/20$. The reliability diagram of each classifier in Figure 2.4 (a) is constructed by plotting the mean of predicted values (x-axis) against the fraction of true positive samples (y-axis) of each stack. Calibration plots diagnose the reliability of a predicted value of an event compared to actual relative frequencies and help understand relative calibration levels of different predictive models. Closer curve to diagonal indicates higher model reliability [282, 283]. A well-calibrated classification model will obtain a curve fall near the diagonal line.

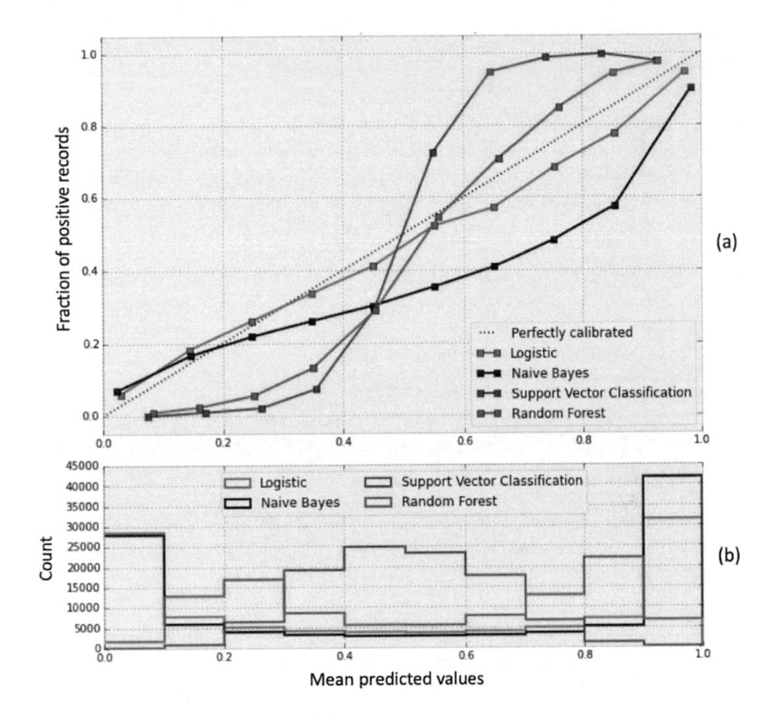

Figure 2.4: Reliability Diagrams: Evaluating and Comparing Classifier Predictions, cf. [191, 203]

As can be seen from the reliability diagrams in Figure 2.4, not every classification algorithm outputs calibrated probabilities directly. LR naturally produces well-calibrated probability estimates, as it directly optimizes log-loss. A coordinate point at $(0.45, 0.41)$ on the LR curve reveals that when the LR model's predictive confidence of the positive class is 0.45, about 41% of the samples are in fact positive. NB is an over-confident classifier and tends to push the prediction score toward 0 or 1 because it assumes that features are conditionally independent of each other. While in RF, prediction scores close to 0 or 1 are rare but frequently appear around the positions of 0.2 and 0.9. The SVM curve reveals a characteristic sigmoid-shaped distortion and pushes the posterior probability scores away from 0 and 1 because it identifies the optimal solution by maximizing margins. This demonstrates that SVM and RF are under-confident when predicting samples that do not belong to the positive class and over-confident otherwise since the uncalibrated probability scores are biased [48]. There are mainly three reasons for biased probability estimation. First, the classification model's assumption may not match the actual data distribution (e.g., the assumption of independence among predictors in NB). Second, valid features have not been used in the classification model.

Third, the training algorithm is insufficient, and the solution obtained by convergence is not the optimal solution for the prediction cases.

Probability calibration is a post-processing operation applied after the trained classification model determines the prediction scores of samples. Models involving the concept of probability calibration provide advantages in performance improvements by transforming the derived classification estimates into a more meaningful probability space [22, 283]. Embedding the probability calibration module into the classifier fusion method optimizes the probability estimation of classification algorithms and enables the computer-aided diagnostic system to better harmonize the predicted values of multiple models in order to further support probability-based diagnosis [44].

There are two commonly used approaches for rescaling and normalizing the classification scores to probabilities distributed in the interval $[0, 1]$: Platt scaling and isotonic regression [191]. Using a sequence of synthetic data points as an example (blue scatter line in Figure 2.5), the x-axis represents the mean predicted values, and the y-axis represents the fraction of true positives. The black dotted reference line appears along the diagonal yield a perfectly calibrated relationship $y = x$. The red dashed curves demonstrate the calibration plot with fitted probabilities using Platt scaling (a) and isotonic regression (b) [123, 169].

Platt scaling, also called sigmoid or logistic calibration, is a parametric approach that applies a logistic regression model to transform the predicted scores from trained classifiers for the purpose of reducing the log-loss error metric [36, 206]. It is suitable for models that have sigmoid curved outputs (e.g., RF or SVM) and small data sets. Isotonic regression is a non-parametric approach that uses a weighted least-squares regression model to transform probabilities [283]. A monotonicity constraint fits the data to a segmented non-decreasing constant function in which ordering plays an important role. The isotonic regression algorithm flexibly slices a sequence of observations and performs stepwise fitting to preserve or reverse a given data sequence [62, 128].

One of the simple iterative algorithms for solving the isotone optimization problems is the pool adjacent violators algorithm (PAVA) [78]. Since the isotonic regression can be formulated as a convex function programming problem with linear constraints, the PAVA calibrates posterior probabilities by applying a ranking method to the classifier's estimated scores (subjective probabilities) obtained by predicting test instances [62, 108]. Depending on the monotonicity constraint, the sequence of data points can be divided into different subsets. The PAVA approach traverses the entire sequence to find the violations of the constraint [123]. For a data point that violates the restriction, PAVA applies linear regression to adjust the score into the most suitable value based on monotonicity [28]. Sometimes it is required to modify the previous scores to ensure that the new data point does not break the constraints [283]. As a result, each score will get one calibrated probability value, and each point will be at least as high as the previous point on the calibration curve (red dashed curve in Figure 2.5(b)) [245].

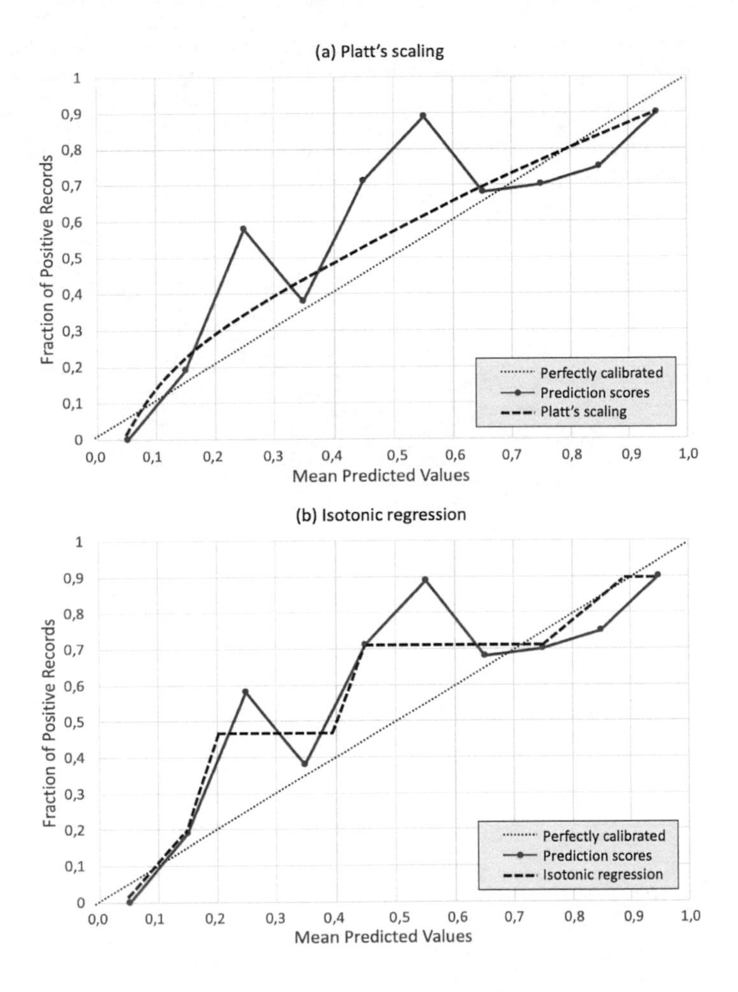

Figure 2.5: Probability Calibration Techniques: Comparison of Platt Scaling (a) and Isotonic Regression (b) [123, 169]

Niculescu-Mizil et al. [191] examined the prediction results from 10 supervised learning algorithms and the effectiveness of Platt scaling and isotonic regression for calibrating the predicted probabilities. For example, Figure 2.6 shows the effects of two calibration methods on SVM compared with LR. The research results show that Platt scaling is particularly effective for max-margin methods, such as SVMs and boosted related algorithms. However, it has less influence on well-calibrated models such as logistic regression and neural networks. Isotonic regression can correct monotonous distortion, but it is easy to overfit on small data sets. Calibration leads to more accurate predicted probabilities for decision-making. Since calibration is a monotonic transformation, it does not significantly affect model prediction accuracy measures, such as ROC, AUC, precision, and recall [203]. It is recommended to use isotonic regression for calibrating large data sets because it does not require the training process, and it is more efficient than Platt scaling [245].

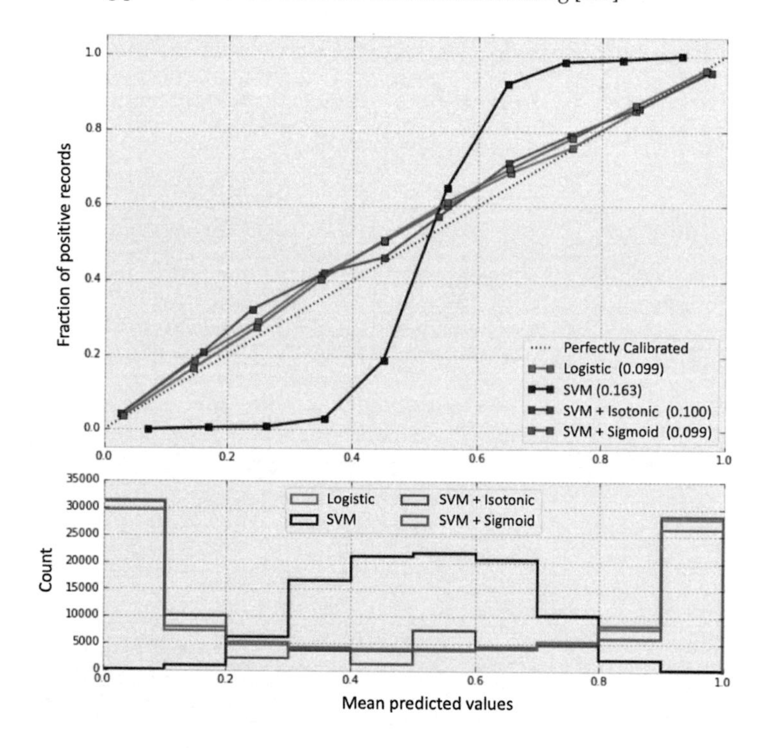

Figure 2.6: Assessing the Effectiveness of Platt Scaling and Isotonic Regression in Calibrating SVM Classifier, cf. [203]

To use either Platt scaling or isotonic regression to obtain a valid calibration model and avoid introducing unnecessary bias, the data set used for calibration should be independent of the classification model's training set. The reason is that taking the prediction results of the classification model to train the calibration model is equivalent to retraining the model with the same data set, which can lead to overfitting. Therefore, in probability calibration operations, cross-validation is often involved in disjointing a training data set into two portions. One portion is used to train the classification model, and the other is used as a validation set to calibrate the predicted values to actual probabilities.

Two indicators are generally used to evaluate the calibration method's effectiveness quantitatively: Brier score and log-loss. The Brier score can be viewed as a cost function that calibrates a set of prediction probabilities. Based on the Brier score's value, the diagnostic system decides whether to support the initial prediction result. The lower the Brier score for a set of prediction values, the better the model's calibration effect [54]. The log-loss function maps the classifier's prediction scores to the posterior probability and quantifies the accuracy of the classifier by penalizing the classification error. Log-loss can be used to evaluate the probabilistic output of the classifier and represent the probability distribution [288]. Minimizing log-loss is equivalent to maximizing the accuracy of the classifier. The closer the prediction scores are to the actual probability, the smaller the log-loss value. In many scenarios, especially multi-classification tasks, log-loss is suitable for evaluating the assurance level of a classification model in predicting instances belonging to each class [283].

The probability calibration approach can improve the comparability between the predicted values of the classifier [124], which enables decision-makers to observe the prediction ranking of different classifiers under the same criteria, thereby enhancing the accuracy of the final decision. The empirical results from Niculescu-Mizil et al. [191] show that after probability calibration, boosted trees, RF, and SVMs predict the best probabilities. To calibrate an ensemble of classifiers, Zhong et al. [288] proposed a refined construction that applied isotonic regression to adjust the combined prediction scores estimated from all selected classifiers. In addition, a manifold regularization is incorporated to smoothen the discontinuities and avoid overfitting. Experimental results show that the proposed approach is more effective than calibrating each classifier individually, especially when the selected classifiers are highly diverse. In a monotone regression procedure, PAVA can decrease the Brier score and cross-entropy and is more robust than the scaling learned by other popular normalization algorithms (e.g., min-max normalization) [6].

2.6 Data Visualization

In the era of information technology, industries have been increasing their emphasis on the importance of data, resulting in increasing demand for integrated solutions of data mining, prediction analysis and consolidation, and visualization of results [251]. On the basis of data management and analysis techniques, data visualization methods extract abstract data attributes and organize them into graphs to display associations, comparisons, and structural relationships [51]. Data visualization utilizes techniques from various disciplines including computer graphics, image processing, computer-aided design, and human-computer interaction technology to visually represent data through expression, modeling, multidimensional display, and animation [52]. Data visualization has diverse applications ranging from simple statistical charts like bar, line, and scatter charts to complex ones like heat maps, bubble charts, tree charts, and dashboards. Additionally, the quality of available development tools, from professional database software to visualization libraries based on various programming languages, is increasing [94, 267].

A sophisticated statistical analysis of data records can bring greater insight. However, the analysis results are usually presented in intensive text and tables, making it difficult to consult and easy to lose focus. Research has shown that the human brain processes visual information faster than textual information [142], and observing data from a visual or physical perspective is more comfortable and more accessible [263]. Compared with textual reports, graphic information provides an at-a-glance interpretation of complex data characteristics and captures data relationships more specifically and persuasively [50, 211]. The purpose of visual data reports is to provide interactive collaboration support, helping professionals gain insights into analysis results, reveal hidden phenomena through impressive visuals, and integrate their professional experience to make better decisions, formulate appropriate strategies, and take action [52, 251].

The advantages of data visualization are conspicuous. It intuitively represents the complex information and regulations in the data and effectively reveals the variation tendency. Through straightforward but meaningful graphics, data visualization simplifies the feature retrieval process during knowledge discovery and reinforces the interpretability and comprehensibility of mining results [52]. Converting abstract concepts into infographics promotes the transparency of information [84], consequently stimulating the knowledge exploration process and facilitating problem-solving and decision-making [254]. The interactive data analysis and exhibition platform not only provides data presentation in multiple aspects, but also highlights information dynamically according to the research targets [84]. Meanwhile, diagrams can provide visual objects for discussion, feedback, and reflection [43]. As each person's knowledge structure differs, the information they glean from diagrams also varies. Data visualization offers a structured method to unveil characteristic patterns of research objects, internal correlations, and the diverse nature underlying phenomena, providing evidence for scientific discovery, engineering development, medical diagnosis, and business decisions.

As a crucial link between clinical diagnosis and scientific research, an interactive visual data analysis module can help medical professionals and domain experts master clinical information, such as identified symptoms of a disease, deviating patterns, curative effects, and the need for attention or improvement. In the research of questionnaire-based RD diagnostic support, the data visualization module targets on presenting orphan disease features and patients' clinical manifestations in order to explore the potential patterns and trends from a large number of medical records and clarify the pathogenesis of specific diseases [174]. With the graphics superiority effect of the data visualization application, medical researchers can collect important disease attributes and group patients' symptoms more efficiently. It makes it easier to integrate the evidence of disease classifications and provides essential elements for final inference [211]. From a long-term perspective, data visualization enables identifying comparative data variations, making preliminary predictions and assumptions. The diagnosis support system's visualization model consolidates experts' experiences into a skill collection, allowing newcomers, especially practitioners, to master prerequisite knowledge rapidly. Furthermore, data sharing between researchers and remote MDs improves studying, diagnosing, monitoring, and treating medical disorders [50].

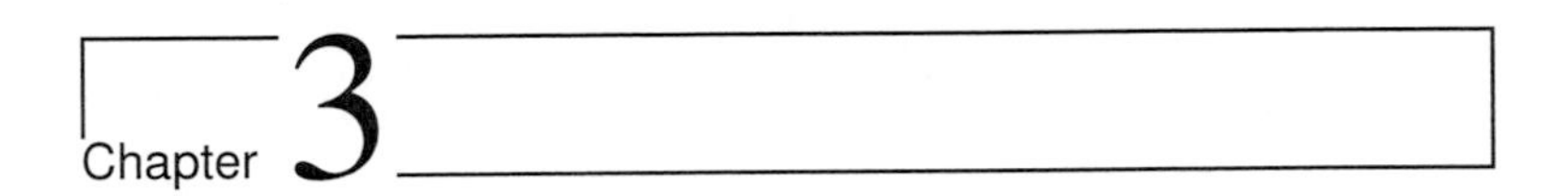

Empirically-Based Disease Prediction Model

A computer-aided disease diagnosis support relies on the precision of RD symptoms collection and the impeccability of medical experts' diagnostics. Due to the data scarcity and dispersity of RDs, the inconspicuous early disease manifestations and data collection difficulties challenged the development of a robust and accurate diagnostic system. As the kernel algorithm of the diagnosis support system, this research is centered on the orphan disease feature extraction and pattern recognition by utilizing patients' experience-based questionnaire data. To improve the diagnostic accuracy of the decision model, the disease prediction model that employs a three-layer safeguard screening mechanism is introduced in this chapter, which involves a statistical classification ensemble, a response similarity evaluation process, and an exceptional situation retrieval phase. This chapter starts with an architectural overview of the designed RD prediction process, followed by the introduction of the working principle of each module in the three-layer screening mechanism. The selection of classification models and the classifiers' combination method are discussed, as is a probability inspection table that provides the performance boost to the target algorithm. In the end, this chapter conducts feasibility experiments using three different data sets to evaluate the performance of the classification model and discusses the accuracy and practicability for future implementation.

3.1 System Diagnostic Procedure

By observing collected symptom descriptive data under the guidance of medical experts' diagnoses, the supervised learning model analyzes the relationship between symptom variables and extends the results to a universal pattern that delineates the characteristics of the disease in order to identify similar unknown cases [5]. According to the experimental results (introduced in Section 3.2), an ensemble of appropriate supervised classification methods was selected to establish the comprehensive diagnostic assistant for optimizing the RD identification process. Figure 3.1 depicts an overview of the diagnostic concept for a new patient.

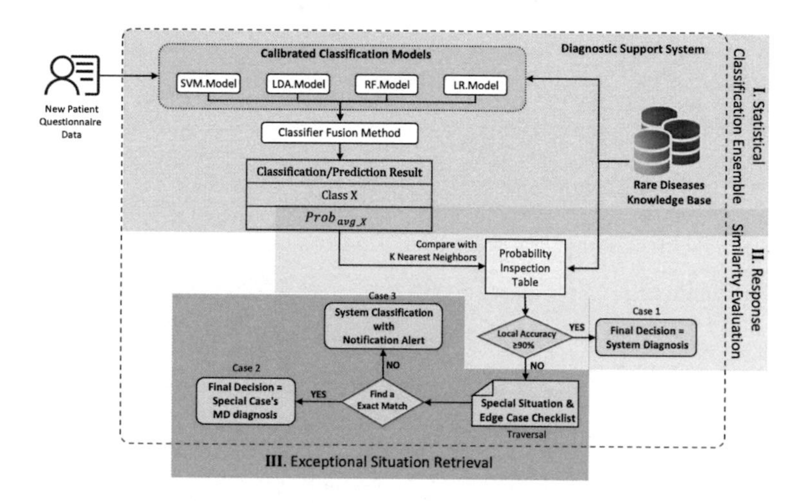

Figure 3.1: Computer-Aided Diagnostic Process: Classifying New Patients Based on Questionnaire Responses

The identified disease patterns and symptom correlations are consolidated into the system's knowledge base for fast invocation. When processing a new patient questionnaire response (as a target case) according to the questionnaire topic, the diagnosis support system invokes the corresponding trained ML models and probability inspection table (introduced in Section 3.4) for cross reference retrieval. To make an elaborated system diagnosis, the system activates the three-layer screening mechanism. The computational core of the RD diagnostic support system is a classifier fusion algorithm that consists of four ML models. In the statistical classification ensemble module, a set of calibrated classification models (SVM, LDA, RF, and LR) make predictions for newly received questionnaire data individually and concurrently. The classification model selection and calibration process is introduced in Section 3.2. Their prediction results are summarized by the classifier fusion method (introduced in Section 3.3) to obtain a comprehensive prediction result with its corresponding probability $(ClassX, Prob_{avg\_X})$.

Inspired by case-based reasoning [137], an alignment and scoring scheme was approached in this study. The idea of case-based reasoning is that MD-diagnosed cases can be reused as reference precedents for solving similar problems and interpreting new situations. For a new pending problem awaiting decisions, a repository that stores previous cases and corresponding MD diagnoses is retrieved to find similar situations. If a source case has the same questionnaire responses as the target case, the related solution for that source case will be referenced as a proposed solution for the new case. Otherwise, the system-suggested solution will be reviewed and adjusted to better fit the new issue. Once the answer given by the system is evaluated as practical, the system will retain the newly learned case and its solution as an update of the resource library — that is, the learning ability of a case-based reasoning methodology [1, 137, 268].

The response similarity evaluation module applies the k-nearest neighbor case retrieval method [268] to constructively support MDs with more rigorous diagnostic evidence. The degree of similarity can be calculated based on their eigenvalues, and a threshold can be determined to converge similar cases in a region. In this study, the threshold was set to **0.05**.

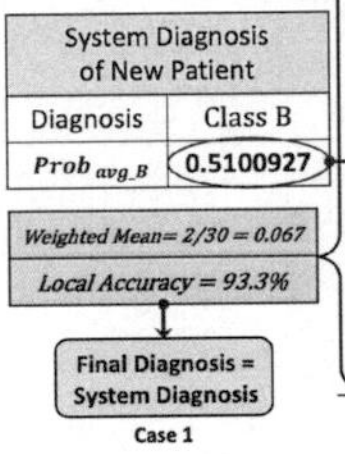

Neighbor i	Weight	Probability Inspection Table				
		Diagnosed Patient ID	Calibrated Prediction Probability (CPP)	System Diagnose Type	Medical Doctor Diagnostic	Error
		...	...	...	...	...
1	1	PID_3524	0.4801339	A	A	0
2	2	PID_2983	0.4801339	A	A	0
3	3	PID_4612	0.4901385	A	A	0
4	4	PID_0087	0.4996841	A	A	0
5	5	PID_5130	**0.5083275**	B	B	0
6	5	PID_1895	**0.5113462**	B	B	0
7	4	PID_7256	0.5139854	B	B	0
8	3	PID_0342	0.5224953	B	B	0
9	2	PID_2509	0.5377491	**B**	**A**	**1**
10	1	PID_0023	0.5584368	B	B	0
		...	...	...	...	...

System Diagnosis of New Patient

Diagnosis	Class B
$Prob_{avg\_B}$	0.5100927

Weighted Mean= 2/30 = 0.067

Local Accuracy = 93.3%

Final Diagnosis = System Diagnosis

Case 1

Figure 3.2: Inspecting Local Diagnostic Accuracy: Probability Inspection Table for the MHH PID Dataset

Using the PID prediction process as an example (the data set information is introduced in Appendix A), Figure 3.2 illustrates the process of comparing the previously acquired new patient's predicted probability score ($Prob_{avg\_X}$) to its k adjacent instances with $Prob_i$ using the PID probability inspection table. In this case, k is an even number depending on the sum of nearby neighbors that are symmetrically scattered on the positive and negative sides of $Prob_{avg\_X}$ that met the following conditions:

$$\{Prob_i \in CPP | Prob_{avg\_X} - 0.05 \leq Prob_i \leq Prob_{avg\_X} + 0.05\} \tag{3.1}$$

where *CPP* is the data collected in the "Calibrated Prediction Probability" column of the probability inspection table. In this example, the new patient is classified by the system as diagnosis $ClassB$ with a prediction probability $Prob_{avg\_B} =$

$\mathbf{0.5100927}$. The position of $Prob_{avg\_B}$ can be located in the probability inspection table, and the value of k can be determined to be $\mathbf{10}$. By forward and reverse lookup neighbors that have similar prediction probabilities to the new patient within the interval, these extracted empirical cases with their diagnostic solutions will become the reference basis for determining the local accuracy of the system, which is reflected by reviewing the proportion of cases where the classifier's prediction results are consistent with their physician's decisions.

In the group of k neighbors, a weight allocation method is applied to emphasize the relative degree of importance and the influencing proportion of different neighbors in the overall assessment. The distribution of weights has a significant impact on the computational results. Elements that are closer to the target case have a more significant influence, therefore assigning a higher weight. In contrast, an influencer from a far distance would need a lower weight than the influencer from closer proximity. According to the distance between a neighbor and the new patient's position in the probability inspection table, $k/2$ is assigned as the highest weight to the nearest neighbors of the target case and decreasing sequentially to both sides. As the example shows in Figure 3.2, when $k = 10$, $weight = 5$ represents a relatively critical determinant, and $weight = 1$ indicates a relatively unimportant determinant. The total influence of k neighbors is measured by the weighted arithmetic mean of whether the system made diagnostic errors with these neighbors [177]:

$$WeightedMean_{\_NewPatient} = \frac{\sum_{i=1}^{k}(Weight_i \times Error_i)}{\sum_{i=1}^{k} Weight_i} \tag{3.2}$$

where the $Weight_i$ is the influence degree of the k_{th} neighbor to the new patient. The $Error_i$ is a record of whether the system made a wrong classification diagnosis for the i_{th} source case. If the system's overall prediction is the same category as the MD's diagnosis, the value of $Error_i$ is marked as 0. Otherwise the $Error_i$ is 1. The local accuracy of the classification system among k source cases is determined by the following equation, which represents the credibility of the system prediction within a local interval:

$$LocalAccuracy = (1 - WeightedMean_{\_NewPatient}) \times 100\% \tag{3.3}$$

In the example shown in Figure 3.2, within the region of ten consecutive neighbors around $Prob_{avg\_B}$, each neighbor is assigned a weight value. The nearest two neighbors have a $5/30 \times 100\% = 16.6\%$ probability of influencing the final diagnosis, and the farthest two neighbors have a $1/30 \times 100\% = 3.3\%$ chance of affecting the final decision. In this example, since there is only one instance in position $i = 9$ that has a discrepant system prediction, the local accuracy of the system diagnostic ability is $(1 - 2/30) \times 100\% = 93.3\%$.

According to the system's diagnosis results for new patients and the corresponding local accuracy, a scenario discussion and distribution procedure is triggered after cross-referencing the classification results of k neighbors, and the discussion result will be regarded as the output decision of the system. If the statistical comparison results within the local range exhibit a strong diagnosis precision ($LocalAccuracy \geq 90\%$), the system's prediction result is selected as the final decision (*Case 1*). In the example mentioned before, the local accuracy ($\mathbf{93.3\%}$) is greater than $\mathbf{90\%}$. Therefore the situation fits *Case 1*, and the final diagnosis of the new patient is the system prediction. If the system makes an unconfident diagnosis ($LocalAccuracy < 90\%$), a special situation and edge case checklist is introduced for the exceptional situation retrieval process. The inventory recorded all questionnaire responses that the system could not diagnose correctly. In this procedure, the new patient's responses are successively compared to the particular cases on the checklist. If an exact match is found, the system will switch its final decision to the MD's diagnosis of the matching case (*Case 2*). If neither of these decisions is possible, the system will make a statement about the debatable diagnosis and push a notification about further investigation suggestions (*Case 3*). In the situation that all classifiers point to one disease category but with a low confidence level, it is possible that the actual diagnosis may be completely different. A prompt message accompanying the system's prediction result indicates which variable has affected the classification and led to the weak diagnosis due to some insufficient evidence. Such prompt messages can assist doctors in examining the relevant symptoms cautiously. Once the patient receives a confirmed diagnosis from the MD, the answer sequence will be recorded on the special situation and edge case checklist for further analysis and discussion.

3.2 Classification Model Selection and Calibration

By analyzing the characteristics of RD questionnaire data collected by MHH medical experts, six eligible ML algorithms were initially selected for the experimental analysis and performance evaluation: support vector machine, random forests, logistic regression, naïve Bayes, linear discriminant analysis, and k-nearest neighbors. In order to integrate multiple ML algorithms with different measurement scales and guarantee a more accurate model evaluation and classifier fusion results, the calibrated-classifier cross-validation(CCCV) method was introduced to convert the prediction scores of different classifiers to actual probabilities.

Figure 3.3 depicts the process of applying k-fold cross-validation (in this research, $k = 10$) to train and calibrate a binary classification model. In the CCCV method, the labeled data were shuffled randomly and split into $\mathbf{10}$ groups. In each cross-validation iteration, the method sequentially choose the ith group ($1 \leqslant i \leqslant k, i \in \mathbb{N}$) as the hold-out validation set for fitting the calibrator. The remaining $\mathbf{9}$ groups, as the result of $k - 1$, are used to train a duplication of the selected classification model. The k-fold cross-validation approach ensures that unbiased data are used to fit the calibrator in each iteration. The probability calibration method (introduced in Section 2.5) transforms the trained classifier's predicted scores of the ith validation data set into the likelihood of an event occurring. In general, Platt scaling is selected if the model's calibration curve has sigmoid characteristics and

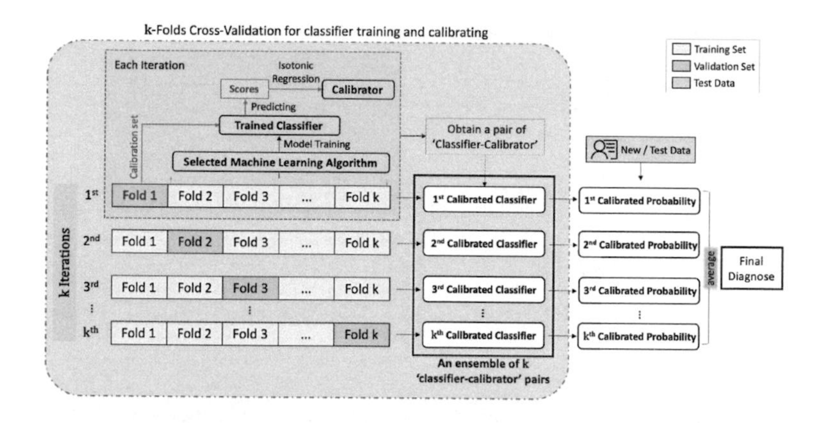

Figure 3.3: Calibrated Classifiers Training with K-Fold Cross-Validation

the amount of data is limited. Platt scaling is an efficient approach when the uncalibrated classification model is under-confident and the predicted output of each binary class has a normal distribution with similar variances (symmetric calibration error) [203]. If a large amount of data is available for probabilistic calibration, isotonic regression is the preferred choice, as it is more effective at non-sigmoid calibration curves but tends to overfit on small data sets [245, 283]. However, the research of Tabacof et al. [249] shows that isotonic regression has better calibration performance and a more flexible range of applications than Platt scaling. Isotonic regression was selected in this study because it is appropriate for calibrating any monotonic distortion without pre-inspecting linear or non-linear models, and can provide more accurate probability estimates for intermediate samples.

The CCCV procedure described in Figure 3.3 will obtain a series of trained classifier-calibrators (k pairs) for an ML algorithm, where each calibrator maps the prediction scores of its corresponding classifier. When predicting a new patient's questionnaire response, each classifier-calibrator pair of the iteration will provide a corresponding diagnosis according to its learning results, respectively. The final probability for each binary classification label X ($Prob\_X$) is acquired by averaging the k prediction results of the calibrated classifiers collection, computed by the following formula:

$$Prob\_X = \frac{1}{k}\sum_{i=1}^{k} Prob(i^{th}\ \text{Calibrated Classifier}, ClassX) \tag{3.4}$$

The ultimate diagnostic result of the new coming patient based on the comprehensive assessment of the ML model is the class label that acquires the highest calibrated probability.

In order to select efficient learning models for the classifier fusion operation, the LOOCV method (introduced in Section 2.3.1) was applied to evaluate the performance of six classifiers (SVM, RF, LR, NB, LDA, and KNN), since this procedure can use as much data as possible to participate in the model training process. A subset of the collected MHH questionnaire data focusing on distinguishing rare disorders was used for model evaluation, which included **354** samples with complete responses and corresponding diagnostic labels. Among this subset (namely "RD versus NRD"), **234** records were diagnosed as RDs, while another **120** were diagnosed as non-rare diseases (NRDs), the detailed description can be found in Appendix A and the reference paper [93]. The LOOCV procedure assesses the model quality by consecutively holding one sample out from the data set as a test sample. Each learning algorithm uses the remaining **353** records for 10-fold CCCV model training, and the calibrated classifier makes independent predictions on the hold-out test sample. Therefore, after **354** times LOOCV iteration, each data record is extracted once as the test data to obtain a final class label prediction with an average calibrated probability value (calculated by Equation 3.4) from each of the trained classifiers. The overall accuracy of each classification model is calculated as the proportion of all system diagnoses that are consistent with MDs' decisions in the data set.

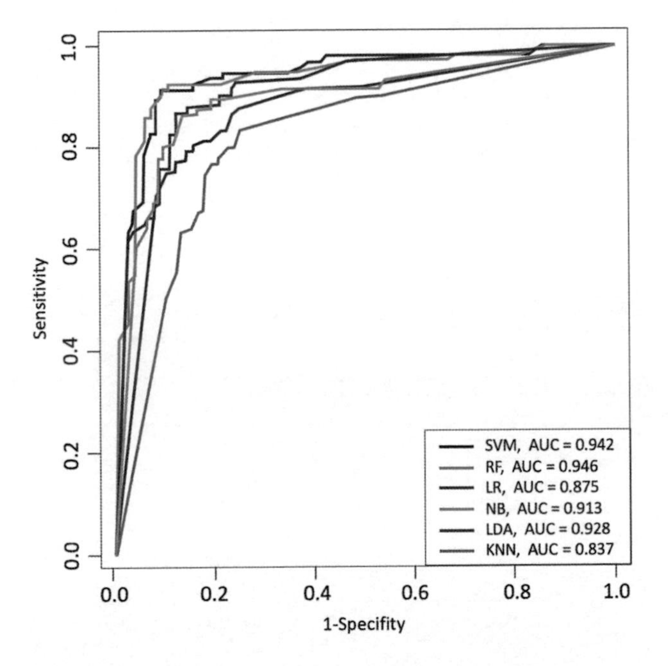

Figure 3.4: ROC Curves and AUC Values: Comparing the Performance of SVM, LR, LDA, RF, KNN, and NB Models using the MHH RD Dataset

The ROC curves and AUC were used as the evaluation indicators to compare the diagnostic ability of six ML classifiers. The ROC curves (introduced in Section 2.3) plot the correlation between *TPR* (or *Sensitivity*) and *FPR* (or 1− *Specificity*) at different classification thresholds. The corresponding AUC provides a visual representation of the aggregate measure of each classifier's performance. The higher the value of the AUC, the better the prediction accuracy of the classifier. As depicted in Figure 3.4, both the RF and SVM models predict the class labels efficiently, while the KNN model underperforms. The model validation process helps find a balance between model performance and intelligibility and selects the appropriate ML algorithms for a classifier ensemble.

3.3 Ensemble Method for Binary Classification Tasks

In the previous section, a set of base classifiers were trained and evaluated with the data set "RD versus NRD". Since different base learners have different strengths and expressiveness on the same data set [85], the accuracy, diversity, and independence of selected classifiers need to be investigated when integrating an ensemble of ML models [227]. The composition of classifier fusion is determined by examining the comprehensive diagnosis precision and computational costs of different combinations [149]. After experimenting, four well-performed ML algorithms were selected as the RD diagnosis support system's core components: SVM, LDA, RF, and LR. As an essential step in multi-classifiers system, the classification model ensemble method is illustrated in Figure 3.5.

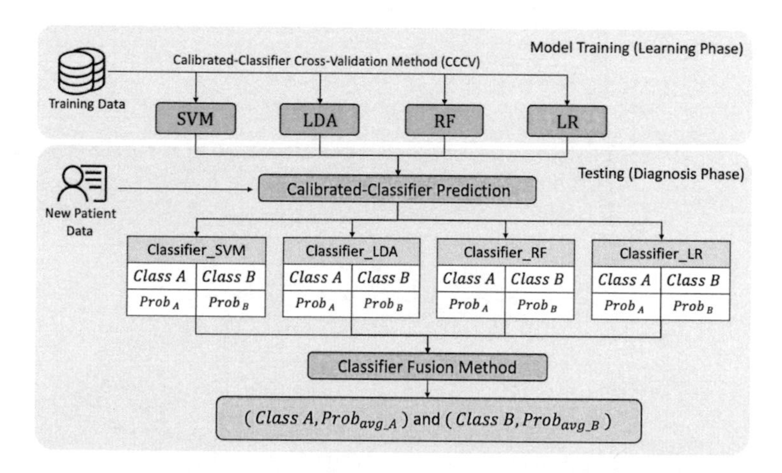

Figure 3.5: Enhancing Binary Classification Performance with Ensemble Learning

The ML-based disease diagnostic task is a two-stage process. During the learning phase, each classification algorithm is trained independently through the CCCV method. In the diagnosis phase, each calibrated classifier predicts the new input case individually. The new patient questionnaire data will therefore obtain four independent sets of diagnostic inferences and corresponding probability values, denoted as $\boldsymbol{Classifier\_j(ClassX, Prob_X)}$, where $\boldsymbol{X}$ represents the binary classification label $\boldsymbol{A}$ or $\boldsymbol{B}$, and $\boldsymbol{j}$ denotes the four different classification models. The $\boldsymbol{Prob_X}$ of the $\boldsymbol{Classifier\_j}$, as the calibrated probability score, represents the confidence of the model $\boldsymbol{Classifier\_j}$ in predicting the new sample into $\boldsymbol{ClassX}$. The classifier fusion method aggregates the calibrated prediction results of SVM, LDA, RF, and LR algorithms for each class label by corresponding probability values, denoted as $\boldsymbol{Prob_{avg\_X}}$:

$$\boldsymbol{Prob_{avg\_X} = \frac{1}{4}\sum_{j=1}^{4} Prob_X(Classifier\_j)} \tag{3.5}$$

In the ensemble of four classifiers, each category receives an average probability score $\boldsymbol{Prob_{avg\_X}}$ as a representative evaluation indicator, and the final diagnosis of the classifier fusion method is the class label with the highest probability. If $\boldsymbol{Classifier\_j}$ is more certain about a diagnosis, it will obtain a higher $\boldsymbol{Prob_X}$, and its prediction will have more impact on the final decision of the classifier fusion method.

$$\boldsymbol{Diagnosis = Class(max\{Prob_{avg\_A}, Prob_{avg\_B}\})} \quad \text{(in binary classification)} \tag{3.6}$$

The classifier fusion method takes advantage of multiple individual learners by evaluating their precision and derives the optimal decision from the combined probabilistic scores. In contrast to the conventional classification method that interprets patients' diseases with an ML algorithm, the calibrated classifier ensemble method improves the interpretability of RD symptoms. The ML-based diagnostic system improves its performance and accuracy by accumulating expert experiences and previously diagnosed patients' symptom records. Continuously gathering reference data from diagnosed patients allows the computer-aided disease prediction system to retrain and re-evaluate.

3.4 Probability Inspection Table for Local Validation

A local reference retrieval procedure was proposed to support physicians in providing more rigorous diagnostic evidence. For the purpose of local validation, a probability inspection table was generated for each questionnaire subject as an indexed catalog of empirical instances sorted by predicted probability value. The program compares the system's prediction probability for the new patient with similarly predicted cases in the repository and evaluates the classification results by referring to the system's diagnostic accuracy within the local range. Screening locally comparable source cases prompts doctors to possibly different diagnosis conclusions — especially for specific RDs that may occur — and inspires them to observe patients' symptoms from different perspectives.

Figure 3.6 shows the process of generating a probability inspection table. The 10-fold cross-validation is applied to the multi-classifiers ensemble method to traverse diagnosed patients' questionnaire data. In each cross-validation iteration, the reference data are randomly scrambled and proportioned into **10** folds. Among them, nine folds (**90%**) of data are re-aggregated and rearranged for CCCV model training (mentioned in Section 3.2). Therefore, the four members ($1 \leq j \leq 4$) of the classifier fusion method will experienced 10 iterations ($1 \leq i \leq 10$) to obtain a collection of classifier-calibrator pairs.

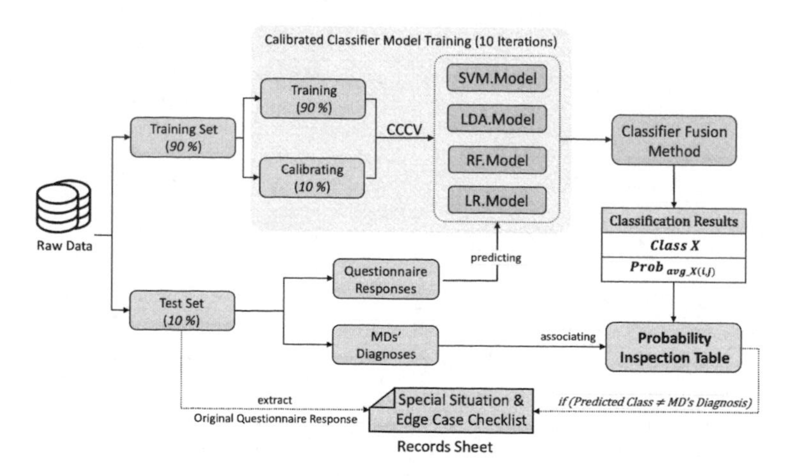

Figure 3.6: The Process of Deriving a Probability Inspection Table

After obtaining 4×10 calibrated classifiers, the remaining one fold of raw data (**10%**) are used as test data. Their questionnaire responses part is applied to each calibrated classifier to obtain the system predictions ($\boldsymbol{ClassX}$) and corresponding probability values ($\boldsymbol{Prob_{avg\_X(i,j)}}$). Then, the MD's diagnoses records carried by the test data itself is used to judge the correctness of the system's diagnostic results. The $\boldsymbol{Prob_{avg\_X(i,j)}}$ for test data is calculated in Equation 3.7:

$$\boldsymbol{Prob_{avg\_X(i,j)}} = \frac{1}{4}\sum_{j=1}^{4}[(\frac{1}{10}\sum_{i=1}^{10}\boldsymbol{Prob_{\_X,i}}), \boldsymbol{Classifier\_j}] \tag{3.7}$$

The iteration continued until all of the data were taken as test data to obtain their assortative system diagnoses. The probability inspection table was generated by associating each instance's original MD's diagnosis label with the corresponding system's prediction, which contains five elements: the ID of the diagnosed patients, their system diagnosis type, the corresponding calibrated probability sorted in ascending order, their MD's actual diagnostic result, and an error flag to mark whether the system predictions are consistent with the doctors' diagnoses. The sequentially arranged calibrated prediction probability values are considered to measure the similarity of patients' questionnaire responses — an application example is mentioned in Figure 3.2. If there is a discrepancy between the system's prediction and the MD's decision, such an instance is extracted into a special situation and edge case record sheet, and the corresponding error flag is recorded as 1. By collecting questionnaire responses for which the system had difficulties assertorically diagnosing the patient's condition or the classification results are inconsistent with the MD's decision, a record of special situations and edge cases assists the system in adjusting diagnosis bias by screening for particular case that have the same structure as the new patient's questionnaire record.

3.5 Validation of the Classification Model

In order to validate whether the designed disease identification and classification approach meet the medical considerations and physicians' expectations in practical applications, the LOOCV method was applied in the validation process to determine the reliability and generalizability of the system through the provision of objective evidence. The overall accuracy of the classification model is calculated as the proportion of all system diagnoses that are consistent with MDs' decisions. It can corroborate whether the system is able to accomplish its desired purpose in the intended operational environment.

Table 3.1: Assessing the Diagnostic Accuracy of the Calibrated Classifier Ensemble Method across RD, PID, and BC Datasets

Data Set		Size	Classifier Fusion	Calibrated Classifier Ensemble Method
MHH Data	Rare Disease Subset (RD vs. NRD)	354 (234:120)	78.3%	**83.4%**
MHH Data	Primary Immunodeficiency Disorder (PID)	126 (64:62)	85.3%	**88.9%**
UCI Machine Learning Repository	Breast Cancer (BC)	569 (357:212)	96.1%	**97.1%**

Three data sets were used to validate the classification model's performance: an RD subset with **354** valid records (RD versus NRD) [93], a PID data set with **126** samples [185], and a breast cancer (BC) data set from the open access UCI machine learning repository that involves **569** instances [70, 170] — detailed information about these three data sets can be found in Appendix A. Table 3.1 summarizes the validation results and compares the simple classifier fusion method with the calibrated-classifier ensemble model designed in this research. The rightmost column records the diagnostic accuracy of the three-layer safeguard screening mechanism mentioned in Section 3.1. As a comparison, the accuracy of the simple fusion method was measured and discussed. As an attempt at the beginning of this research, the originally applied multi-classifiers fusion model does not have the enhancement process of local accuracy verification and screening mechanism.

The experimental validation results demonstrate that the implementation of the calibrated classifier ensemble method with local probability comparison significantly improves the diagnosis accuracy. The precision of diagnosing RDs improved from **78.3%** to **83.4%**, and the diagnosing ability of PID increased from **85.3%** to **88.9%**. The accuracy of diagnosing BC only increased by one percentage point because the classifier fusion method already made a reliable diagnosis.

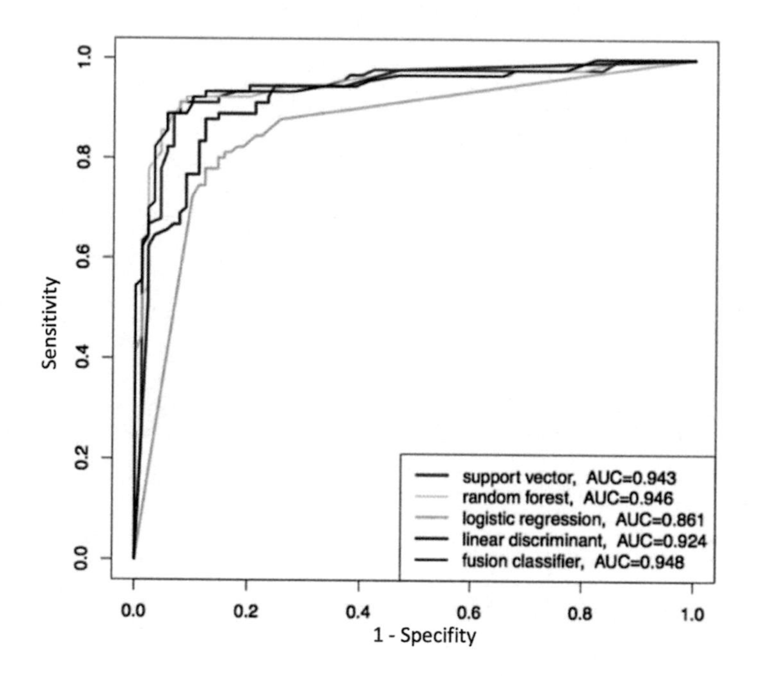

Figure 3.7: Comparing Fusion Method to Individual Classifiers: Performance Evaluation through ROC Curves and AUC Values [93]

Grigull et al. [93] analyzed the diagnostic sensitivity of different classifiers by using a subset of RD questionnaire data (rare pulmonary hypertension and cystic fibrosis versus common non-rare disease) for evaluating the predictive performance of the multi-classifiers fusion method and its member classifiers (introduced in Section 3.3). Figure 3.7 presents the ROC curves and the AUC values of different classifiers. As can be observed from the graphic, the classifier fusion method reached the AUC value of $\mathbf{0.948}$, which outperforms other selected single classifiers. The validation results revealed that the classifier fusion recognition method combines multiple ML algorithms' advantages in obtaining powerful predictive and interpretation capabilities. The proposed three-layer screening mechanism for RD prediction presents substantial performance improvements over any single classifier in supervised modeling.

3.6 Feasibility of the Disease Classification Model

This chapter introduced a supervised ML-based calibrated classification model for the assistance of questionnaire-based RD diagnostics. In contrast to objectively ascertainable information (e.g., disease symptoms or laboratory inspection reports), subjective information (e.g., physician experiences, perceptions, and opinions) is often not directly tangible or storable. Especially for the diagnosis experience of RDs, the difficulty in collecting experienced information has resulted in low accessibility and reusability. The RD-related diagnosis information requires qualified acquisition, analysis, and integration techniques, and it needs to be appropriately managed in a knowledge base.

The system's applicability, operability, and feasibility analysis were deployed based on periodical interviews and workshops with MHH physicians. This feasibility analysis aimed to determine whether the prediction results of the computer-aided diagnostic system can support MDs in diagnosing early-stage RDs while investigating the applicable conditions and limitations. The theoretical and practical results of this study enable to positively answer $\boldsymbol{RQ}1$. As a disease diagnostic assistant for symptom pattern recognition and potential patient prediction, the proposed classification model provides a thorough investigation mechanism for adequate diagnosis using limited resources. Effectively analysis of questionnaire records of diagnosed patients collect disease symptoms comprehensively and aggregate valuable diagnosis recommendations.

The ensemble of diagnostic results from calibrated learning models derives optimal estimation results by utilizing each classifier's advantages, which reduces classification gaps and increases the transparency of the inference procedure. The probability inspection table provides the possibility of cross-referencing the diagnosed patients within a specified range and consolidates different MDs' experiences into an independent skill repository. The local probability comparison process combined with the calibrated-classifier fusion method effectively sorts different cases into reliable statistical predictions and weak predictions. Such a structure helps correct system deviation and optimizes the error distribution of the prediction model.

It is particularly helpful in filtering out poor diagnostic outcomes when the patients' symptom trends make it difficult to decide between two diagnostic types. This approach provides a substantial foundation of evidence for a more precise and reliable comprehensive inference and also provides data support for the study of individual variation interpretation of RD symptoms.

However, the system has certain limitations. For instance, the ML-based system diagnostic accuracy is highly correlated to the amount of collected reference data. Predictive systems with high reliability are built on sturdy data integration, but the currently received data are still relatively insufficient. Although LOOCV validation results confirmed the disease prediction accuracy of the calibrated-classifier ensemble method, the lack of exploration time and real pending cases to be diagnosed hinder the practicability and effectiveness assessment of the module. Thus, there is a capacity shortage in actively operating the module into the developing and testing phases. In addition, collaborating with different medical institutions has a significant impact on the analytical investigation of RDs. In order to maintain and continuously improve the performance of the disease classification and prediction model, it is crucial to periodically trace the updates of clinical medical diagnostic progress in registered patients. Once a doctor's diagnosis that contradicts the system's classification result is received, the patient's questionnaire responses and confirmed doctor's decision would be recorded for future model training.

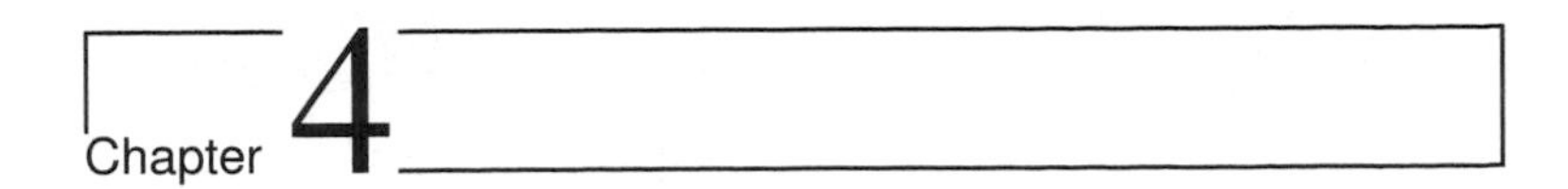

Interactive Visual Inspection of Research Findings

Medical diagnoses and consecutive treatment recommendations require a persuasive theoretical foundation, often aligned with laboratory tests and experts' conclusive interpretations, respectively, to the individual's symptoms [138]. From a data-centered perspective, applying statistical analysis and ML algorithms can surpass human cognition and knowledge reserves to determine the complex interrelations of RD symptoms. However, the ML-based prediction results often lack an explanatory format for interpreting the derived causality and reasoning procedures [94, 251]. The scarcity of relevant medical explanations for some RD questionnaire answer patterns and their symptom relevance hinders practitioners from understanding the scientific discoveries and mechanisms for diagnosing orphan diseases. To investigate possible methods that assist physicians in better understanding the associations between the symptoms of target orphan diseases and the diagnostic result from the model-driven classification algorithms, ***RQ*2**, defined in Section 1.2.2, is the research subject of this chapter.

A supplementary visualization concept is proposed to observe disease-related questionnaire data from multiple analytical dimensions. Combined with complementary data analysis methods, the implemented perspectives include questionnaire response pattern observation, the distribution of diagnostic accuracy rates of different classifiers, analysis of the impact of specific question responses on system diagnosis, and the correlations between different question variables. The intuitive textual descriptions of disease symptoms have high complexity, making it difficult to obtain valuable correlation information efficiently, while the visualization of these variable relations ensures the interpretability and discoverability of complex information that might otherwise be overlooked or unrecognized [84, 251]. Compared to data reports, the visual transformation of data better exposes and highlights the diagnostic rules that are difficult to depict through numerical information to achieve the purpose of more scientific and accurate information interaction [43].

4.1 Symptom Patterns of Different Rare Disease Groups

Confronted with ever-accumulating medical data resources, it is essential to grasp the disease pattern recognition results promptly to assist medical practitioners in diagnosing RDs efficiently. Questionnaire assessment and symptom-related response pattern analysis can infer the probability of a patient suffering from a specific type of RD. The characteristic of an RD can be summarized by analyzing the complex associations between questions. The classification models and disease pattern recognition of the computer-aided diagnostic support system are effective ways to expand the disclosure of these internal dependencies. As the intercorrelation between scientific research and clinical medicine, an interactive visualization application can assist medical professionals in understanding the achievements of the diagnostic system according to patients' medical history, diagnosis results, and therapeutic plans [174]. Studies have shown that for understanding and learning tasks, presenting data and information in the form of graphics is advantageous [4, 159]. Convert abstract concepts into explicable graphical information enhances learners' comprehension of contents and is easy to remember [16, 233, 263].

4.1.1 Forming a Graphical Data Representation

The focus of MDs is to distinguish the response patterns of patients with different orphan diseases and identify how well the response profile of a designated patient matches the general answer patterns of patients who share the same condition. In the process of understanding the distribution of symptom assessment data, it is worthwhile to visualize the characteristics of patient response patterns associated with specific orphan diseases. Particularly, whether the symptom-related questions involved in the questionnaire are equally important for diagnosing an RD, whether some combination of symptoms are the essential indicators to judging an RD, or which symptom-related questions are irrelevant to specific RD, are the focus of medical professionals.

In enhancing the understanding of RD-interrelated symptoms, general practitioners seek an approach that can highlight the answer pattern where all patients under this category are significantly inclined toward a particular option. Question responses that show uniform distribution are not considered the corresponding disease characteristic. In contrast, questions for which most patients choose the same option will attract doctors' attention unless patients from other diagnosis categories also select the same option. In order to intensify those question responses that have a strong correlation with a specified diagnosis group, a data visualization auxiliary process involving sequence logo and p-value was applied to make the disease answer pattern more prominent. The concepts of stacked bar chart in graph theory [33, 270] and sequence logo from bio-informatics [232] were introduced to deliver visual representations of data differences across multiple variables and the value distribution of each variable. A stacked bar chart delivers to general practitioners the absolute or relative frequencies of each possible answer option in the questionnaire and the variation between different questions conveniently.

Using a neuromuscular disease data set that contains **46** questions and **366** data records as an example (data set introduction in Appendix A), Figure 4.1 shows the questionnaire responses distribution of the Morbus Pompe disease category. In the bar chart, the four different colors represent the four possible answer options of each question, representing the negative to positive strength of respondents' agreement or feelings regarding the question. The height of the colored rectangles denotes the number of patients who chose the corresponding option.

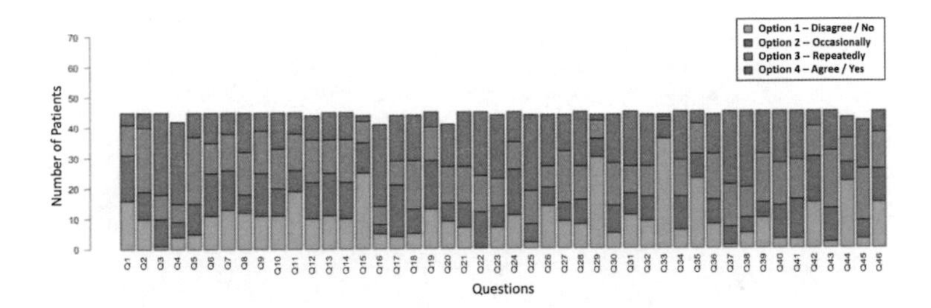

Figure 4.1: Visualizing the Distribution of Questionnaire Data about the Morbus Pompe Disease Group

As can be observed from the diagram, the notable variables in this category are questions $Q3$ and $Q4$, to which patients tended to choose Option 4 (Agree/Yes) and questions $Q29$ and $Q33$, to which Option 1 (Disagree/No) was favored. However, since the y-axis indicates the number of participants and since all accumulated stacks on each question position have the same height, it is tedious to identify these remarkable variables. Although visualization techniques based on multidimensional scaling were introduced to discover outliers and explain how well different diseases can be distinguished [134], these visualizations are too general and provide only an overall view without specific focuses.

A sequence logo is a graphical representation of conserved patterns in DNA or RNA nucleotide sequences or amino acids patterns in protein sequences [232]. A sequence logo was created to summarize the characteristics and information content of aligned sequences, such as consensus and diversity. As shown in Figure 4.2, at a designated position in the sequence with multiple samples, it is of interest to determine whether the same nucleotide always occurs or whether there is high variance. The sequence logo presents the distribution of nucleotides in a DNA molecule using the symbols of four base types (A, C, T, G) stacked on top of each other at specific sites in the alignment and describes whether the nucleotides are variable or conservative. It provides a more detailed view of an alignment than the simple consensus sequence. In the sequence logo, the height of a symbol stack and the relative height of each symbol within the stack are defined by their corresponding Shannon entropy [58, 232]. Entropy is a measure of the unpredictability of an outcome [161]. The greater the entropy value, the higher the uncertainty of the outcome; the lower the entropy, the better the conservation is at that position. The height of the cylinder is negatively correlated with the value of entropy [65].

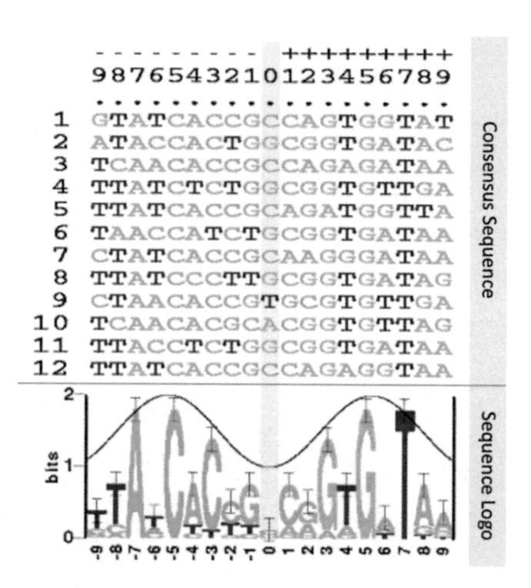

Figure 4.2: Using Sequence Logo for DNA or RNA Nucleotide Sequences, cf. [58]

Regarding the visualization of disease-related questionnaire data, the cumulative frequency distribution plot using the sequence logo technique can deduce important properties and significant variables of the data set through the column's overall height on the vertical axis. The values that were repeated the most can be inspected from individual bricks. The questions correspond to the sequence, and each color represents one possible answer to a question. With an appropriate modification, possible answer options represented by colored bricks will replace nucleotide symbols. Therefore, the overall height of each stack in a sequence logo — denoted by $R_{sequence}(Q_j)$ — depicts the information content or conservation of the sequence at that position [58], which defines the difference between the maximum possible entropy and the entropy distribution of the observed question Q_j. The entropy for each answer to the question Q_j is calculated by

$$H(Q_j) = -\sum_i p_{i,j} \cdot \log_2 p_{i,j} (Q_j = \{Option_{1,j}, \ldots, Option_{i,j}\}.) \quad (4.1)$$

Consequently, the overall height of each stack $R_{sequence}(Q_j)$ is calculated by

$$\begin{aligned} R_{sequence}(Q_j) &= \log_2 N - (H(Q_j) + e_n) \\ &= \log_2 N - \left(-\sum_i p_{i,j} \cdot \log_2 p_{i,j} + e_n \right) \end{aligned} \quad (4.2)$$

In this formula, $p_{i,j}$ is the relative frequency of answer option i of the observed question j, and N is the number of distinct options for the given question. In the case of the designed questionnaire, the maximum sequence conservation per question was $\log_2 4 = 2$ bits, and a small correction factor e_n was introduced (similar to a Laplace correction) for the case when no one chose an answer. The relative height of individual bricks within the stack (denoted by $h_{i,j}$) reflects the information content or significance of the corresponding option in the answer sequence. Such a height is determined by multiplying the frequency $p_{i,j}$ with the total information $R_{sequence}(Q_j)$ at the corresponding column.

$$
\begin{aligned}
h_{i,j} &= p_{i,j} \cdot R_{sequence}(Q_j) \\
&= p_{i,j} \cdot \Big[\log_2 N - \big(H(Q_i) + e_n\big)\Big] \\
&= p_{i,j} \cdot \left[\log_2 N - \left(-\sum_i p_{i,j} \cdot \log_2 p_{i,j} + e_n\right)\right]
\end{aligned}
\tag{4.3}
$$

By applying the sequence logo properties to the stacked bar chart of questionnaire data, a visual improvement to Figure 4.1 is realized in Figure 4.3.

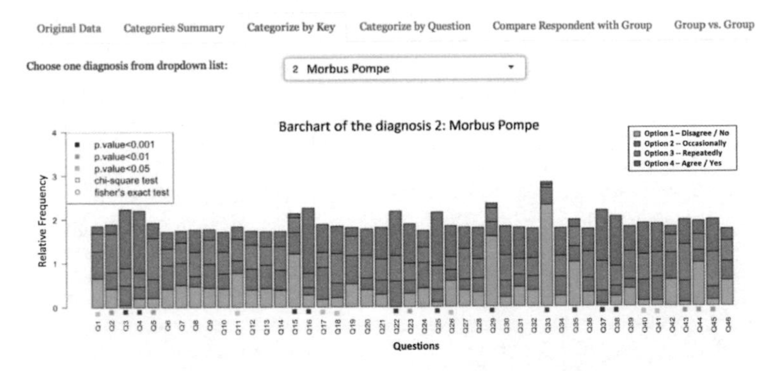

Figure 4.3: Visualizing Morbus Pompe Questionnaire Data with Sequence Logo Technique: Depicting Question Conservativeness and Significance

The relative height of the colored rectangle located on each position of the x-axis no longer equals the number of participants who chose the option but indicated the frequency of represented options that appeared in the question responses sequence. Different colored rectangles under the same question are scaled according to the frequency of corresponding options to depict the consensus or diversity of the answer sequences of such a question. The overall height of the entire stack reflects the degree of conservativeness of each question. In Figure 4.3, the high peaks in the stacked bar chart (e.g., $Q3$, $Q4$, $Q29$, and $Q33$) indicate that almost all participants with Morbus Pompe disease tended to choose the same answer option for the particular questions, which shows high consistency among the answer selections.

Additionally, a set of colored dots was added to the bar chart's bottom to distinguish the interval where the p-value is located. The p-value is the foundation to examine the correlation between the independent variable (symptom-related question) and the dependent variable (diagnosis category), as well as to determine the significance of the results in hypothesis tests [24, 47]. In this experimental study, the null hypothesis (H_0) was set as "there are no relations between the question and the system output diagnosis of a specific rare disease ". Suppose the H_0 statement is true. The p-value is a verification probability that is compared with a pre-specified threshold to determine whether the observed evidence supports or contradicts H_0. If the p-value is less than the threshold (the default p-value threshold for statistical significance is 0.05 [24]), then H_0 can be rejected, and it can be inferred that the alternative hypothesis (H_1 — there exists a cause-and-effect relation) is true [24, 31]. In this stacked bar chart, the green dot stands for a p-value lower than 0.05, yellow means a p-value lower than 0.01, and red dots alert p-values are less than 0.001.

Considering the fact that the data collections in specific RD categories are often limited in size, two complementary test methods were applied to calculate the p-value in this research: Pearson's chi-square test [175] and Fisher's exact test [217]. Both tests are used for determining if the answer pattern significantly deviates from a uniform distribution or whether the answer modes of the two diagnoses are distinctly different from each other but oriented to different data sizes. The Pearson's chi-square test is an asymptotic test using crosstabulation to test the association between different variables and identify differences between observed and expected values [175]. It is advantageous because it is robust and simple to compute, derives detailed information, and incorporates the flexibility of handling data from two categories to multiple categories. Under the same degree of freedom, the larger the chi-square, the more significant the correlation between the independent variables and the dependent variable and the smaller the p-value. However, the chi-square test has some limitations. For instance, the calculation requires a considerable sample size of how each observed category performs. The test becomes invalid if the expected frequency is lower than five [175]. Because of the Pearson's chi-square test sample size limitation, Fisher's exact test was applied when the sample size $n < 40$ and the theoretical frequency $T < 5$ in any of the considered categories. Fisher's exact test is practically applied only in analyzing small samples and gives the exact p-value [217].

As can be observed from Figure 4.3, the question positions with high peaks ($Q3$, $Q4$, $Q29$ and $Q33$) are marked in red, which confirms that these questions are highly correlated with the symptoms of Morbus Pompe disease. Moreover, some potentially overlooked symptom-related questions (such as $Q11$, $Q23$, etc.) are marked in yellow or green depending on the strength of their association with Morbus Pompe disease. These colored dots make experimental data more intuitive and help determine the significant deviation from a uniform distribution of the corresponding question in estimation. As a significant marker for each symptom-related question, the visualized p-value helps MDs and analysts locate the most representative symptoms under a particular RD category rapidly. The combination form of crucial variables associated with the final diagnosis also becomes easy to observe.

4.1.2 Multidimensional Data Characterization

Using the above mentioned strategy to process the collected questionnaire data, the system's visualization module realized the function of strengthening and highlighting the salient features of identified respondent groups, as well as multidimensional data observation. For instance, it supports a holistic view of the answer distribution of different diagnoses in a designated question (*Categorize by Question*), provides observation and comparison of the differences in diversified RD-related answer patterns (*Group vs. Group*), matches the questionnaire answer sequence of a single patient with the gallery of disease answer patterns (*Compare Respondent with Group*), and visualizes possible dependencies among observed variables. Using the aforementioned neuromuscular diseases data set as an example (Appendix A), the following enumerates the functions implemented by the visualization module.

(1) **The Distribution of Patient Answers to a Designated Question**

In order to analyze the response distribution to a single question from different categories, the *Categorize by Question* functional tag visually compares the sensations and experiences of diagnosed patients regarding a designated question. An example is given in Figure 4.4. By selecting a target variable

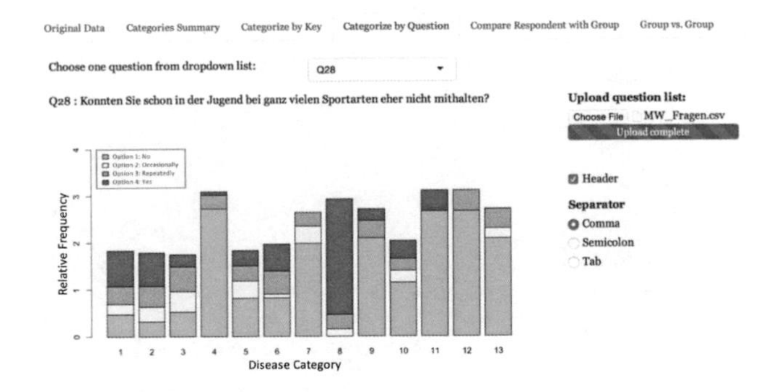

Figure 4.4: Visualizing the Information Content of Variable Q28 across 13 Diagnostic Groups

Q28, a frequency distribution of the neuromuscular disease data set that involves **13** different categories is presented. The graphic illustrates that participants in diagnosis *category 8* predominantly selected *Option 4 (Yes)*, whereas most other groups preferred *Option 1 (No)*. Therefore, ***Q*28** was identified as an indicator to distinguish diagnosis *category 8* from the others.

(2) **Mapping Individual Data into Answer Patterns**

The purpose of mapping a particular patient's questionnaire feedback into an assigned categorical answer pattern is to investigate whether the individual's responses match the general symptoms of a specific disease type. Using a patient questionnaire responses (ID: *MW_P160*) from neuromuscular disease data set as an example, Figure 4.5 shows the result of mapping the patient questionnaire data to their system-diagnosed disease group (Morbus Pompe).

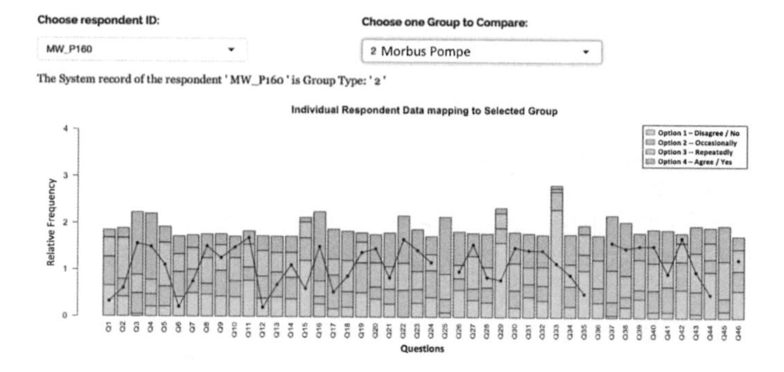

Figure 4.5: Observing Consistency: Patient Questionnaire Data Matches the Answer Pattern of Morbus Pompe Disease Group

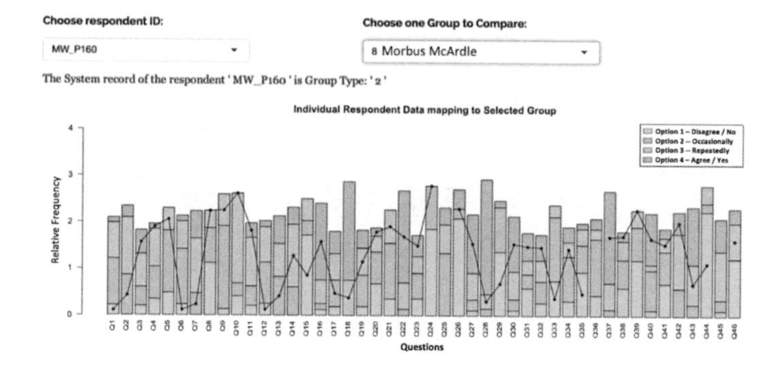

Figure 4.6: Visualizing Incongruities: Patient Questionnaire Data Misaligned with the Answer Pattern in Morbus McArdle Disease Group

The visualization function uses points and lines superimposed on the stacked bar chart to demonstrate the answer sequence of an individual participant. Each point is positioned in the middle of a colored rectangle corresponding to the selected answer option to trace the exactitude degree of the system in diagnosing that patient. It can be seen from Figure 4.5 that the patient's responses to most of the questions were consistent with the answer pattern characteristic of the corresponding system diagnosis.

As a contrast, the questionnaire data from the same patient can be mapped to any other diagnosis category (e.g., diagnosis *category 8*, as shown in Figure 4.6). The graphic presents an apparent discrepancy between the individual responses and the answer pattern of diagnosis *category 8*, indicating that the patient does not meet the criteria for diagnosis *category 8*. The comparison between the stacked bar charts in Figure 4.5 and Figure 4.6 reveals that the patient's answer sequence is more suitable with diagnosis *category 2* than diagnosis *category 8*. The diagram demonstrates the classification model's strategy for diagnosing new patients. It seeks the best-fitting answer pattern as the foundation of system diagnosis.

(3) **Comparison of Response Patterns for Different Diseases**

In addition to providing effective means of verifying individual questionnaire data, the visualization module is also furnished with the pattern comparison function. It supports users in selecting any two different diagnosis groups' response patterns and visually compares their differences and commonalities. For example, Figure 4.7 illustrates the sequential answer pattern comparison between diagnosis *category 1* and diagnosis *category 11*.

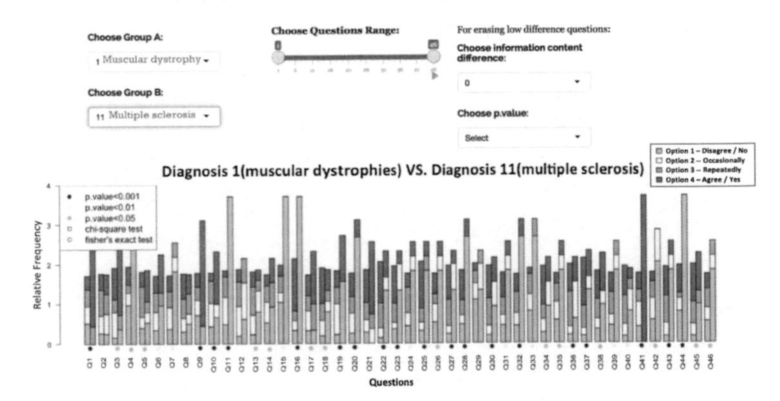

Figure 4.7: Comparing Diagnosis Indicators in Two Disease Categories: Using Stacked Bar Charts with Information Entropy

By selecting the target groups from two drop-down lists, the differences in the patients' responses to each question variable and the variance of patients' performance are presented in the form of a stacked bar charts combination. In each question variable along the x-axis, two stacked bars representing two groups of data sources are displayed next to each other (the left bar is diagnosis *category 1*, and the right bar is diagnosis *category 11*), which helps analysts monitor two disease groups intuitively.

The focuses for medical professionals are the high peaks in the stacked bar chart. The most notable case is two high peaks for one question that have opposite primary colors, which means that the answers to the related question are homogeneous within each diagnosis but heterogeneous between the two diagnoses. It is also of interest if there are remarkable gap and height differences between the two comparison bars of the concerned question. Patients in one diagnosis group tend to equalize select answer options, whereas patients show a very heterogeneous answering scheme in the opposite diagnosis group. For more intuitive observation, the user-oriented visualization interface supports selecting restrictions and scope to reduce data dimensionality, eliminate secondary information, and retain and highlight the essential message. As introduced in Figure 4.8, users can manually scale the interval of observed variables and filter out low-difference questions by selecting the information content differences or p-value threshold.

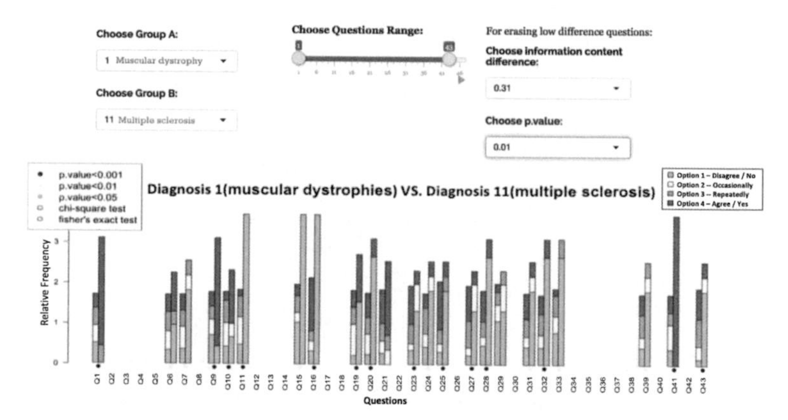

Figure 4.8: Optimize Variable Observation with Customizable Data Analysis: Manually Scale and Filter Significant Questions

The stacked bar chart, along with the analysis of information entropy and p-values, provides an intuitive way to reveal the degree of dispersion and central trends of each diagnosis group's responses to symptom-related answer options. Through the frequency distribution of the samples represented by the graph, the distribution of the population can be roughly estimated, and the differences in patients' response modes between disease groups can be presented, which is helpful for medical experts and scientific researchers in observing data schema that a data table cannot easily represent.

4.2 System Diagnosis Distribution with Heatmaps

Heatmaps use a sequence of color codes to visually reflect the similarities or differences in data. The color gradients defined in heatmaps are used to effectively highlight changes in observed topics and data [87]. The intensity of the color can reflect ranking, frequency, data discrepancies, and so on. Heatmaps are widely used data analysis means in different fields. For example, in bioinformatics, heatmaps are often used to present the analysis results between the experimental and control groups [87]. In the implementation of the questionnaire data visualization concept, the application of heatmaps provides an overview of different classification models' diagnostic results for the same sample case. The diagnostic differences between four classification models and the correction ability of the fusion method for erroneous diagnoses can be easily captured by expressing the system's predicted categories and the corresponding probabilities through color shades. By investigating the changes in the system diagnosis to the minor adjustments of a question's response value, it is possible to grasp the essential factors from the disease symptom sequences that affect the system's judgment and evaluate the stability of the system diagnosis results.

4.2.1 Heatmap of Classifiers' Accuracy Distribution

In order to visually compare the diagnostic certainty and accuracy of the classifiers utilized in this research, the heatmap conspicuously depicts the observational perspectives of the prediction differences between the classifier SVM, RF, LR, LDA, and classifiers fusion method. Using the PID data set that has **126** qualified samples (introduced in Appendix A) as an example, Figure 4.9 shows the classification results for each test sample obtained by each classifier using the LOOCV method. In this heatmap, the first through fifth columns store the diagnostic results obtained from different classification methods. The sixth column (rightmost) marks patients' actual diagnoses confirmed by MDs. Each row represents a test sample, and different color codes exhibit different disease categories. In a binary classification problem, a light blue block indicates that the system has diagnosed the patient with PID, and dark blue denotes that the patient has not been diagnosed with PID by the system. Therefore, each row indicates the classification results of the same patient's answer sequence by different classifiers. The diagram depicts the proportion and distribution of correctly classified samples by comparing the classifier's color blocks with the color representing the accurate diagnosis in the rightmost column. The mispredictions of the corresponding classifier become the most striking part, such as the dark blue blocks in the large light blue area. Moreover, it can be observed in the figure that some classifiers, such as SVM and RF, are more accurate than other classifiers. Still, the classifier fusion method has the highest accuracy. Nevertheless, the classification model cannot diagnose some patient questionnaire data correctly. In this case, it is of great significance to use the probability comparison table to correct the classification deviation.

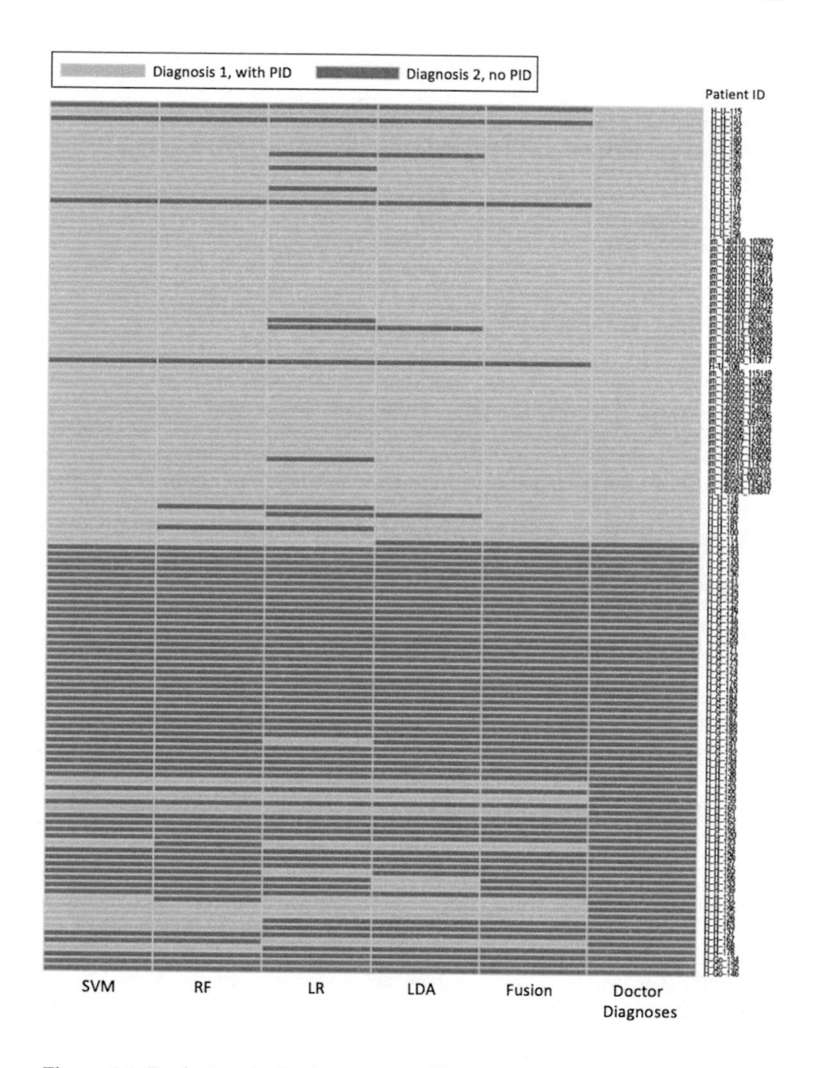

Figure 4.9: Exploring the Performance of Classifiers on the MHH PID Dataset: A Heatmap Visualization of Diagnostic Result Distribution

Figure 4.10 has a one-to-one corresponding relationship with Figure 4.9. In this binary PID classification problem, the color blocks no longer represent the diagnostic result of the classifier, but the calibrated probability value (refer to Section 3.2) for the corresponding patient obtained by each classification model. The magnitude of the color transition represents the probability values.

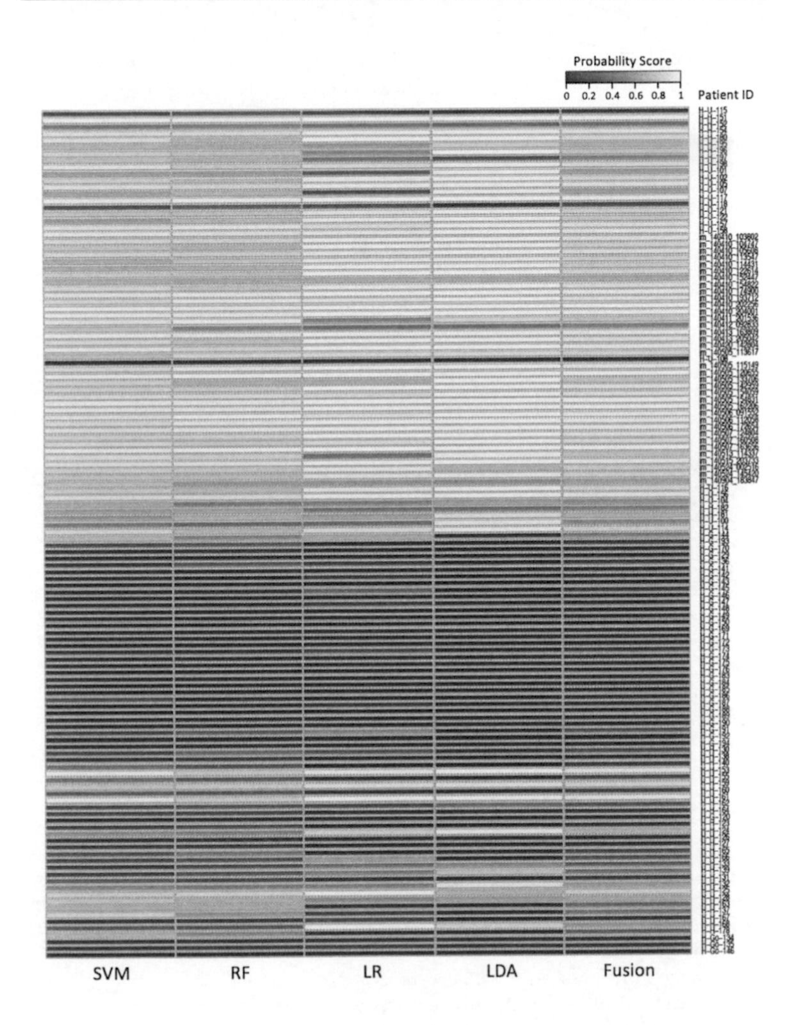

Figure 4.10: Exploring the Performance of Classifiers on the MHH PID Dataset: A Heatmap Visualization of Prediction Probability Distribution

The dark red-colored blocks demonstrate that the probabilities are approaching 0, inducing the system to make a no-PID diagnosis, and the light yellowish-colored blocks indicate the probability values close to 1, which prompt the system to make a PID diagnosis. The intermediate color indicates the uncertainty of the classifier regarding their corresponding diagnoses. Such a graph intuitively reflects the confidence of different classifiers in making a certain diagnosis, and it can be observed that even if the diagnostic confidence of a single classifier is insufficient, the classifier fusion method can enhance the diagnostic accuracy for some patients.

Unlike the assessment metrics used to evaluate the overall performance of a classification model, such as accuracy or sensitivity, heatmaps exhibit the prediction results and confidence levels of different classifiers for individual test data with an intuitive, explicit visual impact. The medical practitioners' shared experiences and feedback confirmed that the use of heatmaps to display the operation results of the diagnostic support system can provide visual threads and help MDs precisely locate information agglomeration patterns or tendencies found during statistical analysis and comparative analysis.

4.2.2 The Impact of a Single Question's Response

Transparent and medically interpretable ML results are essential for disease research and patient prognosis, especially in the medical diagnoses field that requires reliable predictions. Understanding the rationale for predictive models in making confident diagnoses is a complex task for general practitioners. In the data-driven computer-aided diagnostic support system, the categorical diagnosis is determined by the values of a series of question variables provided by the respondent. While slight value changes of a question variable due to respondents' uncertainty or decreased attention may subvert the system's classification result. To assist MDs understands the impact of symptom-related question responses on ML-based diagnosis, the designed interactive visual analytic module provides a dependency analysis module that examines the consequence of input variables on the overall prediction results. The concept was proposed to sequentially analyze the questionnaire data of each patient, and iteratively modify one answer within the range of options to investigate whether the question's answer adjustment influences the diagnosis.

Considering the interactive user interfaces, the corresponding questionnaire responses and the system's original diagnostic result are extracted and presented to MDs in a table by manually selecting the patient ID from the drop-down list. Using a patient's questionnaire responses (patient ID: *im_140411_201536*) from the PID data set as an example, Figure 4.11 demonstrates the test results regarding the question variable influences on the diagnostic system's classification result. The test program sequentially alters the selected answer to each question in the extracted questionnaire, while the answers to all other questions remain unchanged. The system will again diagnose the modified answer sequence through the classification model to monitor whether one step alteration of the question's answer would lead to different system diagnostic results. After adjusting the value of each question variable consecutively and testing the system's prediction result, a summary table containing all of the variables that influenced the system's diagnostic and the relevant minimum adjustment is printed. In the example shown in Figure 4.11, the test caught three questions where a modification of the answer would result in a different classification category. The last column, "Jump", represents the minimum distance between the answer modifications that caused the system's prediction adjustment and the original patient responses to the corresponding questions. Since the answers to most questions are on a scale of **1** to **4**, "Jump" equaling **3** means that the answer must be modified from one ultimate value of the scale to the other to alter the diagnostic system's prediction results.

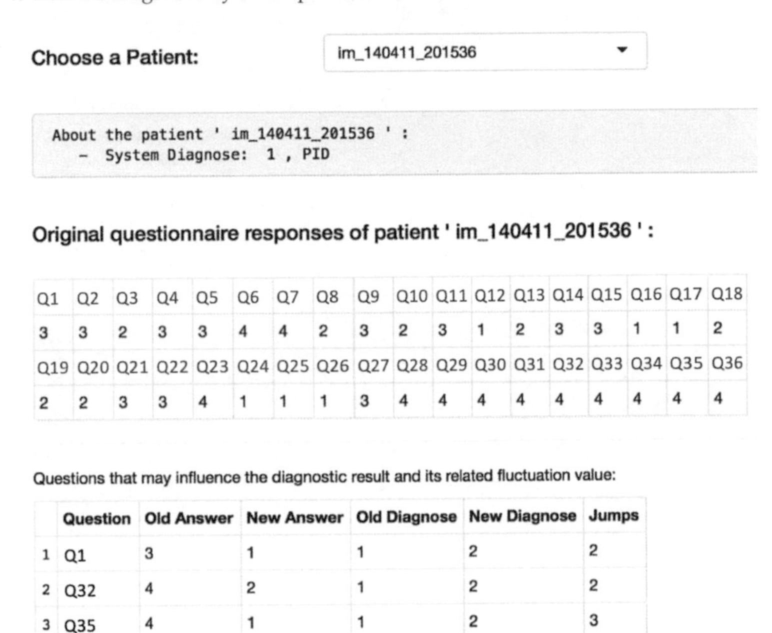

Q1	Q2	Q3	Q4	Q5	Q6	Q7	Q8	Q9	Q10	Q11	Q12	Q13	Q14	Q15	Q16	Q17	Q18
3	3	2	3	3	4	4	2	3	2	3	1	2	3	3	1	1	2
Q19	Q20	Q21	Q22	Q23	Q24	Q25	Q26	Q27	Q28	Q29	Q30	Q31	Q32	Q33	Q34	Q35	Q36
2	2	3	3	4	1	1	1	3	4	4	4	4	4	4	4	4	4

	Question	Old Answer	New Answer	Old Diagnose	New Diagnose	Jumps
1	Q1	3	1	1	2	2
2	Q32	4	2	1	2	2
3	Q35	4	1	1	2	3

Figure 4.11: Assessing the Impact of Answer Adjustment on System Diagnosis: An Example from the MHH PID Dataset

Figure 4.12 visualizes the impact of each question's response on the diagnostic result of the system by applying LOOCV to all questionnaire records in the PID dataset and analyzing the effect of single answer adjustment. The x-axis of the diagram represents the question number, and the y-axis denotes patient ID. Bricks marked with colors indicate that the answer changes of related questions alter the classification model's diagnosis of the corresponding patients. According to the number of options in the questions, there are three levels of option jumps. The red brick represents the "Jump" equals **1**, which marks the most notable variable because a subtle value change to the question would lead to a radically different result. Furthermore, the answers that need to change to an opposite option ("Jump" equals **3**) are marked in green. In the horizontal direction, it can be observed that there are several rows densely containing colored bricks. It illustrates that the responses given by these patients are in an unsteady state. If the answer to any of the marked questions is adjusted, the system diagnosis would be reversed. In the vertical direction, it is evident that question *Q32* is susceptible. Only one jump answer modification of the question could lead to diagnostic prediction changes for most patients (**85.7%**), which is the same as question *Q1* and *Q35*. The columns entirely covered by white blocks indicate that no matter how the answer to the single question changes, it would not affect the final system diagnosis (e.g., questions *Q4* and *Q19*), which are insensitive questions.

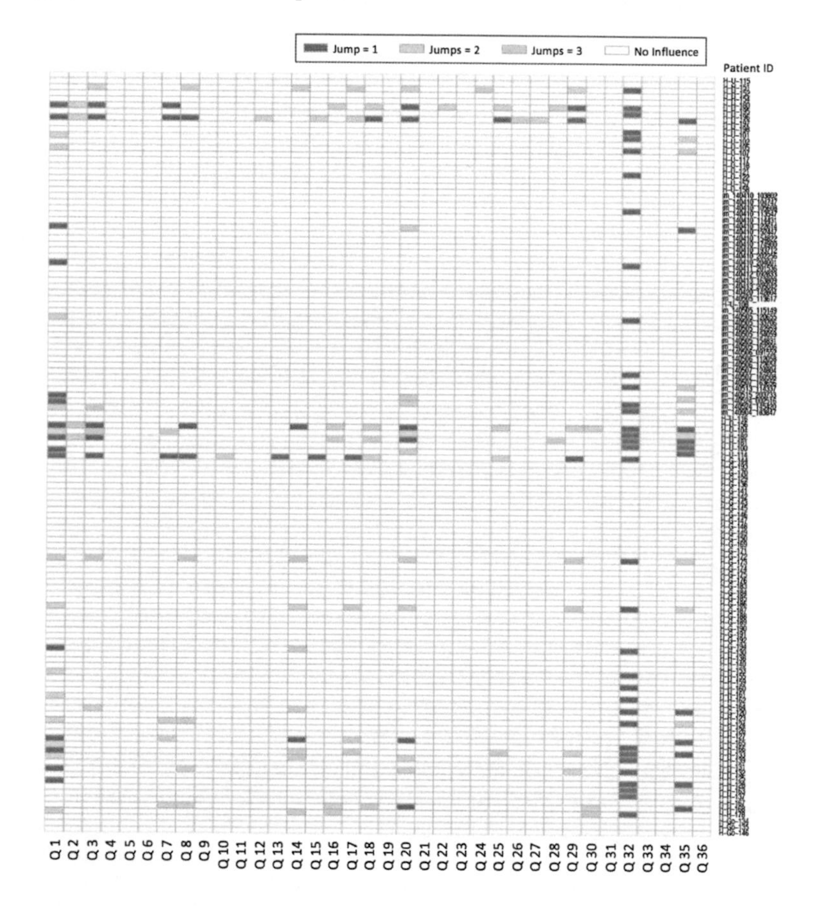

Figure 4.12: Examining the Effect of Single Answer Adjustments on System Diagnosis: Heatmap Visualization of the MHH PID Dataset

The practicality of the implemented data visualization concept was discussed with MHH medical experts during workshops and group discussions. The feedback from practical applications during this study period confirmed that the single variable influences test presented by heatmaps helps medical practitioners understand the impact of a particular symptom on the ML-based system diagnosis. With the exploratory adjustments of questionnaire responses given by the patient and re-diagnosing, the reliability or sensitivity of the system diagnosis can be determined intuitively. For questions where a tiny adjustment to the answer option would significantly impact the system's diagnosis of most patients, an in-depth analysis of their associated symptoms is essential. Conversely, if the adjustment in the answer to a question does not affect the classification results for the majority of patients, it would suggest that the question variable was not a decisive factor in diagnosing the disease. The investigation and visualization of RD-sensitive questions can help MDs grasp disease characteristics efficiently, for better communicating with patients in a targeted manner and screen for potential symptoms.

4.3 Advantages of Diagnostic Data Visualization

For the early diagnosis of RDs, the focus of medical experts is on identifying similar cases by scientifically and systematically summarizing the symptoms of known patients by using data analysis tools and statistical methods. Well-performing ML models have prominent predictive capabilities and high accuracy, but the structural complexity of the model leads to an interpretability reduction [89]. To convert the discriminative results of classification models into easy-to-understand logical relationships, this chapter introduces the concept of interactive data visualization and discusses its feasibility in practice. The process of data analysis can systematically extract information from trivial data sources and search for the rules as the foundation for the statistical decision. Cognitive interpretation of data supported by visualization techniques, such as infographics and optical elements, is a beneficial way to help MDs understand and form an information base that transforms data into knowledge and comprehend the underlying dependencies.

The qualitative feedback from medical specialists demonstrated that the interactive data visualization and analysis tool resolves ***RQ*2**. Visualizing data from different perspectives assists MDs from different analytical backgrounds in associating RD symptoms with the classification model's diagnostic results. The noteworthy specialty of data visualization is that it improves the ability of information interpretation. Compared with traditional ways of presenting data in tables or documents, the visual representation characterizes information more intuitively and makes data more persuasive. Data visualization transforms the information provided by patients' questionnaire data into valuable insights, highlights the significant characteristics of data, and amplifies concerning knowledge. It helps MDs distinguish and understand disease-related symptoms and discover vital factors. Moreover, the subsequent visual representation of symptom dependencies provides valuable support for decision-making and diagnosis scheme formulation.

The observation of disease feature analysis and simulation results using the neuromuscular disease data set and PID data set suggested that different diagnoses have varied peculiarities and symptom patterns, which can be used to distinguish each other. Although the designed questionnaire has a fixed number of questions, the visualization results of the experiment confirmed that the discrimination of a specific RD could be supported by responses to only the questions related to the orphan disease's particular symptoms, which provide a solid basis for the system diagnosis. Under the condition that the crucial information is obtained, the presence or absence of some other questions does not influence the system's final predictions. The accuracy and reliability of the system's classification results are related to the patient's responses to a combination of specific questions. The researchers speculated that directly reducing irrelevant data entry in the form of questionnaires given to new patients can save respondents the time cost of filling out questionnaires while improving diagnostic efficiency, which led to the research of dynamic adaptive questioning method discussed in the next chapter.

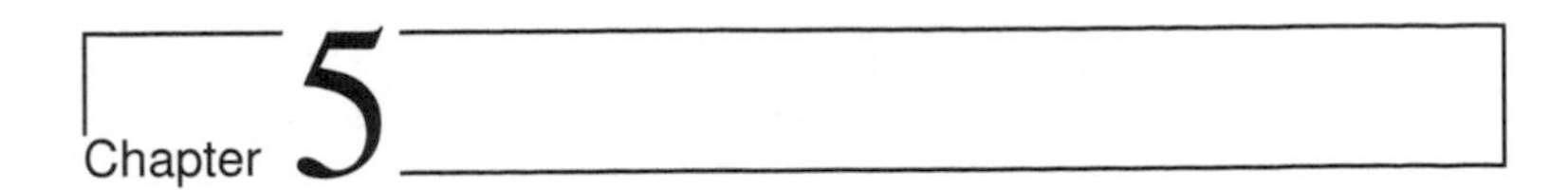

Dynamic Adaptive Decision-Making Mechanism

In the healthcare environment, standardized questionnaires are widely used in the process of collecting patients' subjective perceptions about their health conditions during the diagnosis and treatment journey [113]. These questionnaires usually assess several investigation fields and inquiry aspects focusing on significant symptoms as an appropriate disease indicator. Each symptom-related question provides a different association strength target for diagnosing a specific RD. Some are prominent features of the disease, while others are only recognizable through a combined question set with associated symptoms for a particular disease. In this computer-aided diagnostic support system built upon medical questionnaires aimed at RD screening, the questions were designed as feature variables to measure the symptoms of different diseases.

However, the questionnaires designed by MHH medical experts involve multifarious and tedious disease symptoms, and the permutation and combination of questions imply a hierarchical and progressive investigation [92]. The symptom-related questionnaires considered in this thesis cover multiple rare or non-rare disorders rather than a specific RD characteristic. For example, the collected responses to the RD questionnaire cover several orphan disease categories, non-rare chronic diseases, and psychological disorders. Some characteristic variables are partially or utterly irrelevant to the target disease or have redundancy, while some arguments may correlate with other disease categories. The content of the RD questionnaire can be found in Appendix E.1, and the detailed response rate can be found in the research paper [93]. Patients are often held to fill in extensive and time-consuming surveys to achieve an accurate classification, sometimes including questions that are irrelevant to their proper condition. The number of symptom indicators to be assessed, uncertain answers caused by patients' inattention or distraction, and the interference of irrelevant answers are likely to affect system diagnostics' accuracy. Consequently, when a patient shows the characteristic tendency of a particular disease in the process of answering the questionnaire, it would be highly desirable to centralize the questions to significant indicators that have strong associations with the specific disease. With the purpose of finding a concept to reduce the complexity of the questionnaire according to an individual's condition while ensuring the diagnostic accuracy of the system ($\boldsymbol{RQ3}$), this chapter starts with the examination of correlations between questionnaire size and the system's diagnostic accuracy, followed by an exhaustive searching method for the optimal question combinations in identifying the target disease. A dynamic questioning concept based on ML cognitive construction is therewith proposed to adjust upcoming questions based on the patient's previous responses and realize the questionnaire customization.

5.1 Dimension Reduction Analysis

Patients' experiences during their diagnostic journey have high analytical value and the potential to support physicians considering an RD by using pattern recognition through data mining methods. A classifier fusion model that combines four ML algorithms in distinguishing RDs was introduced in Chapter 3. By analyzing and learning the maximal mutual information from questionnaire data collected from diagnosed patients, the ML methods establish a correlation mapping function from a set of input features to the output discrete classification values. As a diagnostic aid, the disease prediction model helps physicians derive more precise diagnoses by comparing new patients' clinical characteristics with learned categorical information. The experimental conclusions in Chapter 4 proposed that some RDs can be identified and determined by the disease classification model merely through the specific collection of orientated symptom indicators.

In exploratory data analysis, researchers are often interested in identifying promising associations for further investigation and root-cause analysis while filtering out relatively weak referential or prevalent symptoms to reduce the computational cost of modeling and improve the predictive model performance. The research findings and experimental observations of the previous chapters have concluded that the total information content of an RD is obtained from a group of unique features that involve discrimination information about the disease classification. A combination of specific clinical symptom components can collectively indicate a documented disease and lead to recognition by a classification system. By eliminating the weak-tie dependent variables, the number of questions can be reduced, and the classification performance can be improved [46]. From an ML perspective, using irrelevant variables to obtain data that interferes with a diagnosis will lead to poor system generalization capabilities [96]. In order to reduce the number of unimportant eigenvectors and increase the proportion of representative eigenvectors when developing predictive models, it is necessary to understand the impact of individual variables on diagnostic outcomes and quantify the relationship between variables to infer whether they are dependent or independent. The primary objective of this section is to discover complex associations in the high-dimensional RD data set, quantify the degree of interdependence between input and target variables, and further appropriately balance feature dimension and diagnostic accuracy.

5.1.1 Exploring the Interdependence Between Symptoms

For probing incidence relations, there are several parametric or non-parametric methods to quantify the strength and direction (positive or negative) of the relationship between two variables. The principal component analysis (PCA) computes the eigenvectors and eigenvalues of the covariance matrix to identify essential linear correlations [126, 272]. PCA is an unsupervised technique that performs linear transformations on potentially related variables, in order to reduce the dimensionality of large data sets. It reduces the influence of irrelevant variables on high-dimensional data while preserving as much information as possible to make ML more efficient (e.g., image recognition) [76, 101]. On the contrary, statistical-based feature selection algorithms take the target variable into consideration [145]. Variable and feature selection combines an exhaustive searching technique for feature subsets and evaluation metrics to score the input variables' usefulness in predicting the target value and find a subset that minimizes the error rate [60, 96]. However, some correlation measures are restricted to the types of associations that they are able to detect or make assumptions about the underlying distribution of the variables. Since most biological processes are defined by non-linear, complex relationship patterns, experts prefer to use a more comprehensive association method.

To maximize access to the information and potential associations, the exploratory dynamic adaptive questioning concept in this thesis involves conventional descriptive statistics together with the maximal information coefficient (MIC) to determine possible linear or non-linear dependencies [133, 219]. Based on the concept of mutual information (MI), MIC complements the conventional Pearson's correlation coefficient and Spearman's rank correlation coefficient, which analyses linear dependencies and general monotonic relationships in data without investigating non-linear characteristics. The advantages of MIC are generality and equitability, which means that MIC can capture a variety of associations, including linear, non-linear, functional, or non-functional relationships (e.g., cubic, exponential, sinusoidal, superposition of functions) [236]. And the MIC values for different relationship types are similar at the same noise level [133, 219].

MIC measures the degree of association between two variables by assessing the strength of the statistical dependence [219]. Assuming that there is a correlation between two random variables X and Y, the data set $D = (x_i, y_i), i = 1, 2, ..., n$ represents n pairs of corresponding values. A symmetric feature matrix can be created by dividing a two-dimensional grid $G(\tilde{x} \times \tilde{y})$ according to the scatter plot of the two variables, where the two axes are divided into $\tilde{x}$ and $\tilde{y}$ partitions, respectively. The distribution and proportion of these data points that fall into the grid can be regarded as an approximate joint probability distribution and marginal probability distribution[236]. Furthermore, the MIC values between X and Y are obtained by calculating the mutual information of these two variables and normalizing by their minimum joint entropy [220]. Different divisions will result in different probability distribution functions, leading to different mutual information values. In information theory, the mutual information — denoted as $I(X;Y)$ — measures how much information one variable can provide about another variable. It is a non-negative value ($I(X;Y) \geq 0$) that calculated by each variable's entropy (denoted as $H(X)$ or $H(Y)$) and the joint entropy of both variables $H(X, Y)$ [201]. For a random variable, entropy measures the expected average amount of information to transmit an event drawn from the probability distribution [146, 147].

The *entropy* of the two discrete random variable X or Y are:

$$H(X) = -\sum_{i=1}^{n} p(x_i) \log p(x_i) \tag{5.1}$$

$$H(Y) = -\sum_{j=1}^{n} p(y_j) \log p(y_j) \tag{5.2}$$

And the *joint entropy* of variable X and Y is:

$$H(X, Y) = -\sum_{i,j=1}^{n} p(x_i, y_j) \log p(x_i, y_j) \tag{5.3}$$

The *mutual information* between two discrete random variable X and Y can be estimated from the joint distribution according to $p(x, y)$.

$$I(X; Y) = H(X) + H(Y) - H(X, Y) \tag{5.4}$$

The *MIC* value of the two variables X and Y is defined as:

$$MIC(X, Y) = max\{\frac{I(X; Y)}{\log min(\tilde{x}, \tilde{y})}\} \tag{5.5}$$

The algorithm systematically determines the maximum available normalized mutual information on all partitions by considering relationships and their functional properties. Different relation types produce feature matrices with different properties. For instance, strong relationships produce feature matrices with high peaks, and monotonic relationships yield symmetric feature matrices. In contrast, complex relationships yield feature matrices whose peaks are far from the origin. The MIC value, which falls into the domain $[0, 1]$, evaluates how strongly these two variables are correlated through normalized scores and can be interpreted as the percentage of variable Y explained by variable X [219]. $MIC(X, Y) = 0$ represents that there is no statistical correlation between the variables, and $MIC(X, Y) = 1$ means a noise-free relationship of any form [133]. For noiseless functional relationships, MIC is comparable to the coefficient of determination (R^2) [219].

In identifying representative dependencies and strengths of correlations between variables, such as examining the association between patients' symptom indicators and diagnosis from medical specialists, it is essential to investigate the relationships of each variable pair holistically. This section uses a subset of an RD data set that contains **354** patients' questionnaire responses as an example (detailed description can be found in Appendix A and previous work [93]). By analyzing the questions' dependency and the influence of a single question in the system's classification process, MIC measures how each question is associated with a particular RD category and which questions are most helpful in distinguishing between certain orphan diseases. The MIC ranking reveals interdependence between the RD class and the **53** variables (symptom-related questions) involved in the questionnaire. The complete ranking table covering all questions with diagnosis relevance can be found in Appendix C.1. The closer the correlation coefficient value is to zero, the more statistically independent the variables. Vice versa, a coefficient value close to one indicates a strong correlation between the question and the RD group.

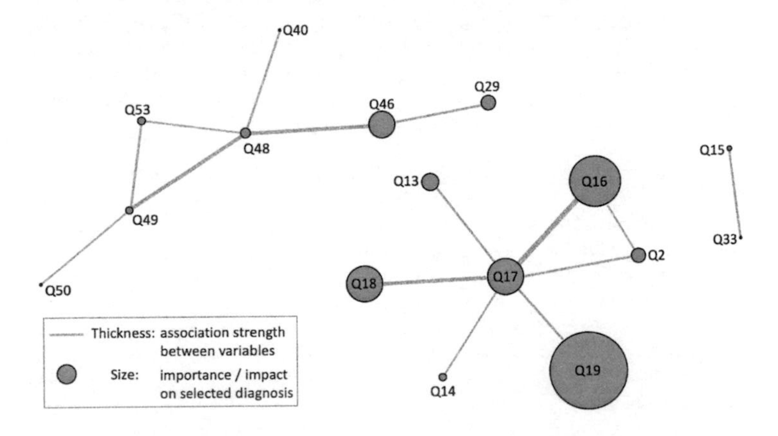

Figure 5.1: Depicting the Dependency Network of RD Questions by Significance and MIC Values using the MHH RD Dataset

In order to visually observe the correlation between different question variables, a bubble chart combined with vectors in the coordinate space was applied to characterize their statistical relationship. Figure 5.1 illustrates the dependency network of the top sixteen critical questions for diagnosing RDs. The diameter of each bubble represents the importance of a question-represented symptom to the selected RD, and the thickness of the line segments indicates the association strength between the conjoint question variables. According to the identified relevance between attributes and disease categories, the ranking table helps promote the targeted evidence collection and integrate fundamental symptoms for the categorical inference. Such supplementary hierarchical graphs visually explain the linear and non-linear dependencies found by statistical analysis, thereby promoting medical researchers' awareness and understanding of the underlying correlation and development tendency of the specific symptoms in RD diagnostics. Discovering the interdependence of comparative data lays the groundwork for questioning mechanisms in dynamic adaptive questionnaire development and directional question selection.

5.1.2 Exhaustive Search for Optimal Question Combinations

There exist various ways to retrench the number of questions within a questionnaire. For instance, statistical significance and the impact or correlation measures can eliminate questions with weak or non-relevant influence on a particular diagnosis. These strategies provide a static reduction of the number of questions to reduce questionnaire time while increasing the system diagnostic accuracy. Dash et al. [60] analyzed **32** different feature selection algorithms in handling various aspects of data and summarized their strengths and weaknesses based on data types and domain characteristics. The research findings revealed that the exhaustive search method (brute-force) is a well-performed technique for knowledge discovery and understanding the dependencies between variables. The exhaustive search systematically enumerates all possible solutions and iteratively examines whether each candidate satisfies the conditions of an optimal solution [184].

In this research, finding a minimal subset of the disease-related questions that can accurately categorize new patients requires an extensive recognition of question-diagnosis relevance. A series of question subsets were generated from the original questionnaire by utilizing the coherence ranking table that examines each question's positive or negative association with a particular disease category. The exhaustive searching process successively removes the lowest prioritized question from the ranking slot and exports the remaining questions, along with the patient's answers, into a new data file. The process traverses all possible question combinations and utilizes the LOOCV method to evaluate the system's capability in classifying patients into the correct diagnostic category while reducing the number of questions. The exhaustive search reveals the trade-off between questionnaire size and diagnostic accuracy of the classification model by delimiting question combinations within the maximum diagnostic association and verifies the feasibility of question combinations according to the system's diagnostic accuracy.

Using the questionnaire that contains **53** questions for RD diagnosis as an example, **354** complete questionnaire responses were used in this experiment. By iteratively removing weakly related questions, **53** sets of questionnaire data with a decreasing number of questions were reconstructed. Figure 5.2 summarizes the brute-force search result of the correlation measurements between the questionnaire size and the system's diagnostic accuracy fluctuations. The horizontal axis of the line chart represents the gradually increased number of questions, and the vertical axis represents the diagnostic accuracy of the classification model.

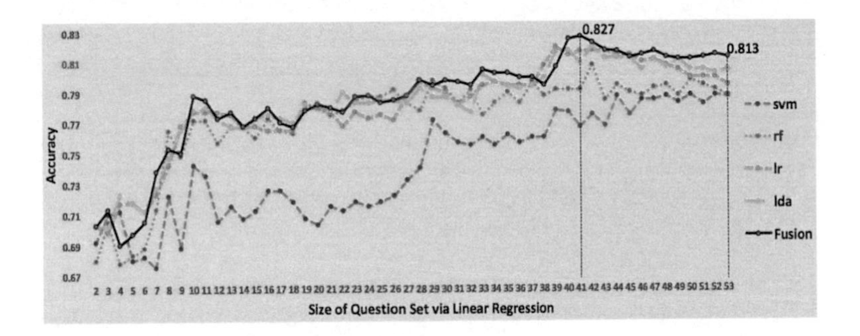

Figure 5.2: Exploring the Correlation between the Number of Questions and Diagnosis Accuracy in Selected Classification Methods

As can be perceived from this diagram, when the questionnaire contains **41** questions, an equilibrium point of the optimal solution is found where the classifier fusion method reaches a diagnostic accuracy of **82.7%**. It is worth mentioning that compared with the original questionnaire with **53** questions, the overall diagnostic accuracy of the classifier fusion method obtained from the reduced-question questionnaire increased by **1.4** percentage points.

Figure 5.2 reflects that the system does not require participants to respond to all questions in the questionnaire to make an accurate diagnosis. The classification model can diagnose new patients by learning from previous cases and efficient qualitative identification. Questions that are beneficial to distinguishing between RDs and NRDs reduce the questionnaire size while supporting faster system diagnostics and higher accuracy. Therefore, a dynamic questionnaire with the capability of selecting questions based on the answers to previous questions is highly desirable. The strategy of a dynamic adaptive investigation survey bridges the gap between maximal possible consideration of all questions and the respect of patients' answers. The purpose is to apply as few questions as possible to collect maximum symptom information and derive a correct system diagnosis.

5.2 Dynamic Adaptive Questioning Concept

The ML-based diagnosis system can assist MDs in classifying and comprehending the symptoms associated with different diseases. Meanwhile, a self-adaptive learning concept was proposed to generate a dynamic questioning mechanism capable of instantly matching the acquired individual symptom indicators with the system-identified disease patterns while filling out the questionnaire. Through the utilization of appropriate data analysis, a dynamic adaptive questionnaire was developed to effectively targets a patient's specific symptoms and yields precise diagnoses, while minimizing the impact of irrelevant variables or noise [162].

The concept of dynamic decision support aims to trace high-probability associated potential symptoms and generate proper decisions under the condition of limited patient-given eigenvalues. In this section, an adaptive channelized question selection mechanism is introduced to realize a customized questioning process with the purpose of RD diagnostics. The mechanism is an iterative inquiry and forensics procedure that continuously collects patient responses to different questions and makes real-time predictions until the classification system can determine the diagnostic category confidently.

The computation is grounded on real-world questionnaire data from diagnosed patients to ensure the adaptability of the dynamic questionnaire. By analyzing a patient's responses to previous questions and correlating the obtained information with known disease patterns, the algorithm utilizes the variable correlation network to infer pending factors that are strongly associated with a diagnosis and issues the next critical question that is the most conducive to the system diagnosis. The seemingly random but purposeful questioning mechanism avoids bias by changing or rearranging the survey questions for each respondent.

Figure 5.3 reveals the adaptive questioning method in the dynamic diagnosis process. The process is comprised of three steps: questionnaire initialization, a new question selection, and diagnostic prediction. At the beginning of the program, a `counter` variable was defined and initialized to 1, in order to count the number of issued questions in a questionnaire that contains n questions. For a binary classification question, the labeled training data collected from diagnosed patients are divided into two isolated groups according to different categories ($\boldsymbol{ClassA}$ and $\boldsymbol{ClassB}$).

Subsequently, the system employs statistical algorithms to recognize data regularities and extract features or properties to describe patients' response patterns to their subordinated categories. Meanwhile, each question's Shannon entropy value is calculated under each diagnostic type to measure the uncertainty of a discrete variable's possible outcomes [235]. The information entropy signifies the randomness of the patients' answer to a question that possesses four options. For example, a question's entropy under one diagnostic category equal to zero denotes that all of the patients tended to choose the same answer. Merging the question's information entropy of both categories measures the orderliness of the patients' responses to the corresponding question. The lower the entropy sum of a question, the stronger the tendency of patients to a specific answer.

In the questionnaire initialization stage, the system assigns the general answer pattern of each disease category to two independent sequences ($\boldsymbol{AS1}$ and $\boldsymbol{AS2}$). The answer sequence is considered the universal answer of a diagnostic class to avoid null values, which will shorten the gap between the questionnaire subset answered by the new patient and the originally completed questionnaire. With the initialization of both categories' universal answer patterns, both assigned sequences would obtain a high probability of their subordinate diagnostic classes, yielding the $\boldsymbol{odds}$ of the two diagnoses $\mathbf{50\% : 50\%}$.

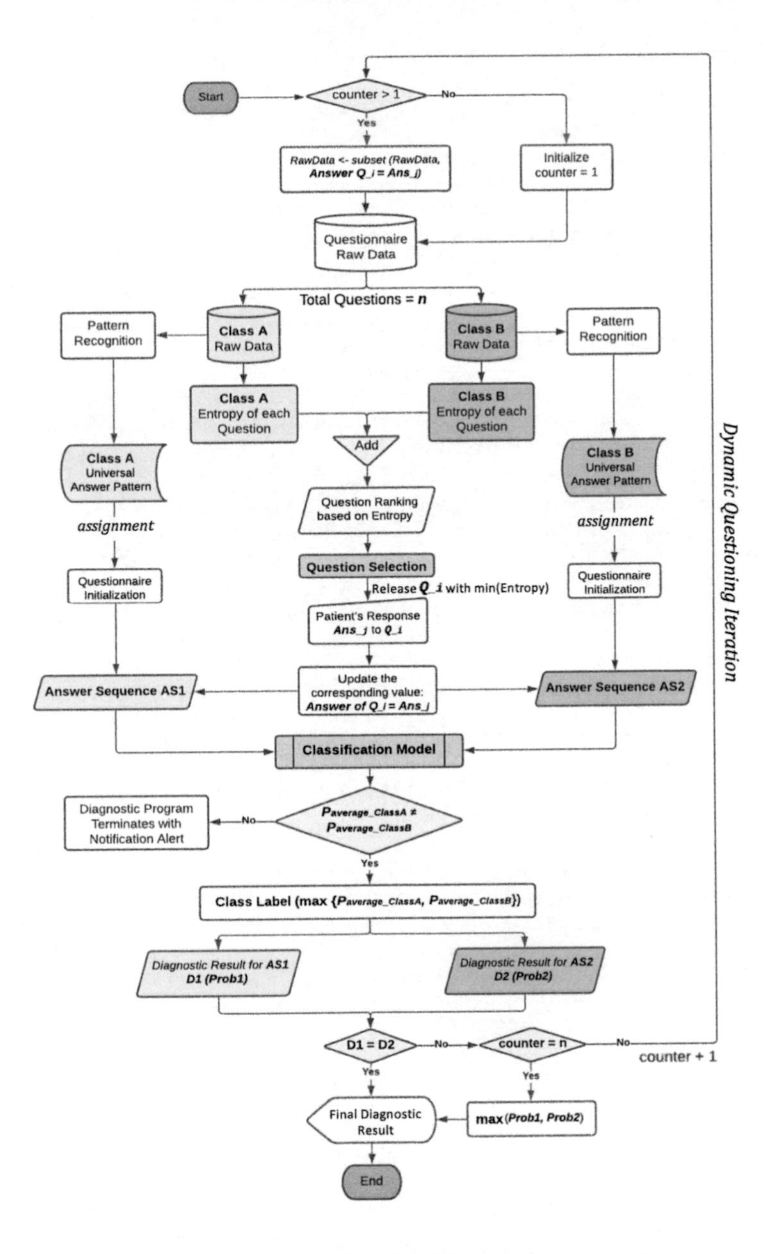

Figure 5.3: Flowchart of the Dynamic Adaptive Questioning Process

In the question selection module, the dynamic questioning mechanism chooses the question with a minimum entropy value from the merged ranking list as the next released question ($Q_{\_i}, i \in [1, n]$). By requesting the patient's response $Ans_{\_j}$ of the question $Q_{\_i}$, the obtained answer is used to update the value of the corresponding question in the answer sequences $AS1$ and $AS2$. That is to say, in each iteration, only the corresponding answer to the system-released question is replaced by the patient response in both sequences, while the answers to questions that have not been referred remain unchanged.

The next step of this process is to apply the calibrated-classifier fusion method (introduced in Chapter 3) to evaluate both progressively enriched answer sequences concurrently and determine the probability that the classification system will vote for each disease category. The assessment of a patient's condition depends on whether the answers provided by the patient match the response pattern of a particular disease group, and the system takes the diagnostic category with the highest probability as the final prediction result.

Hence, the diagnostic result $D1(Prob1)$ and $D2(Prob2)$ for sequences $AS1$ and $AS2$ are acquired, respectively. By comparing the system's confidence level in both diagnosis results, new $odds$ of the two diagnoses are obtained. Under the "optimistic" hypothesis, the patient will answer the continuously released questions in accordance with the answer pattern of a specific diagnostic category, which will cause the $odds$ to deviate from the original ratio. The expected case is that both predictions have the same diagnostic results ($D1 = D2$), which indicates that one of the diagnosis results is deviates from its original one. When the $odds$ sharply depart from $50\% : 50\%$, it means that obtaining the answer to a particular question would significantly clarify the diagnosis. Thus, the system can determine to make the final diagnosis. Otherwise, the system-obtained information from the patient is not enough to make a correct diagnosis, and the next question is requested to be released to collect more evidence.

For each subsequent iteration, a subset of raw data is extracted by filtering out all data with a different value than the patient selection of the current question. According to the $odds$ obtained from the previous iteration and the answer distribution of diagnosed patients in the subset, the dynamic questioning mechanism loops back to the question selection module to recalculate the entropy of the remaining questions and update the corresponding question ranking table. The procedure raises follow-up questions by continually seeking input variables that are strongly correlated with the target variable (diagnosis categories). According to the accumulated symptom indicators, the diagnosis classification model determines the final diagnosis and educes a supportive prediction probability by identifying the two parallel updated answer sequences. Through this iterative approach, the dynamic adaptive questioning mechanism can collect sufficient evidence to deliver a reliable diagnosis while reducing the total number of questions by presenting patient-oriented questions. Suppose that the questioning mechanism has pushed all of the questions in the questionnaire (`counter = n`) but still not obtained a confident diagnosis ($D1 \neq D2$). In that case, the system will select the diagnostic class of the answer sequence that gained a higher probability — $max(Prob1, Prob2)$ — as the final prediction result. A particular circumstance is that the classification model cannot decide on the category of the patient's answer sequence because the obtained probabilities of both classes are equal ($P_{average_{C}lassA} = P_{average_{C}lassB}$). In these circumstances, the system throws an exception, and the diagnostic program terminates with a notification alert. As a retrospective diagnostic support system based on experience accumulation, it is meant to trace the patient's follow-up symptoms and professional diagnosis results. Once a physician's diagnosis contradicts the system's predictions, recording the patient's past questionnaire responses and the specialist's diagnosis can support retraining the learning model and correcting classification bias.

5.3 Validating the Dynamic Adaptive Questioning Method

With the implementation of the LOOCV method, a subset of RD data that contains **53** questions and **354** valid records was used to evaluate the performance of the dynamic adaptive questioning method (the complete questionnaire can be found in Appendix E.1, and RD data set description can be found in Appendix A and related work [93]). In each validation iteration, the LOOCV method sequentially isolates a sequence of individual questionnaire responses from the original data set as the test sequence to simulate the replies from a new patient, denoted as $seq_{\_test}$. The remaining **353** samples are applied to identify the universal response patterns for rare and non-rare diseases. According to the aforementioned dynamic questioning process, the question ranking table computed from the training samples prompts the system to exhibit the appropriate question. And the simulated patient's answer corresponding to the issued question can be located from the sequence $seq_{\_test}$ based on the question ID.

The questioning process constantly selects questions and extracts answers from $seq_{\_test}$ to evaluate whether the classification model obtains sufficient information to diagnose the patient accurately. For the question dispatched in each iteration, the validation process records the question ID and counts the number of questions that the prediction model required to make qualified diagnoses. Table 5.1 demonstrates a fragment of the dynamically formed questionnaire documentation based on continuous diagnosis of gradually enriched patient responses.

It is noticed from the overview of the generated questionnaire subset that the patient's response to a question directly impacts the follow-up questions and the questionnaire size. Some particular symptom-related questions, such as *Q32* — "Have you tried on your own to alleviate your symptoms (for example, by herbal medications, ointments, sports, special diet, etc.)?" — are always issued early to provide substantial evidence for diagnosing the category of a respondent, as they are highly relevant to the diagnosis of orphan diseases. The system can make compelling predictions as the patient's responses gradually match the characteristic pattern of a particular diagnosis type. Some patients answer six to nine questions to receive a final diagnosis, but others require even up to **39** questions to ensure a reliable diagnosis.

Table 5.1: Sample of Personalized Dynamic Adaptive Questionnaire Records for Patient-Oriented Diagnoses

Patient ID	Total Questions	System Prediction	Doctor Diagnosis	Question Record
MW_P018	13	A	A	Q32; Q12; Q26; Q8; Q31; Q20; Q47; Q29; Q21; Q49; Q9; Q7; Q19
MW_P209	39	A	A	Q32; Q12; Q26; Q8; Q31; Q47; Q20; Q29; Q21; Q49; Q9; Q7; Q19; Q14; Q53; ...
MW_P245	13	A	A	Q32; Q12; Q26; Q8; Q31; Q47; Q20; Q29; Q21; Q49; Q9; Q7; Q19
MW_P246	13	A	A	Q32; Q12; Q26; Q8; Q31; Q47; Q20; Q29; Q21; Q49; Q9; Q7; Q19
MW_P247	9	A	A	Q32; Q12; Q26; Q8; Q31; Q47; Q20; Q29; Q21
MW_P001	13	A	A	Q32; Q12; Q26; Q8; Q31; Q47; Q20; Q29; Q21; Q49; Q9; Q7; Q19
MW_P005	9	A	A	Q32; Q12; Q26; Q8; Q31; Q47; Q20; Q29; Q21
MW_P006	13	A	A	Q32; Q12; Q26; Q8; Q31; Q47; Q20; Q29; Q21; Q49; Q9; Q7; Q19
MW_P013	6	A	A	Q32; Q12; Q26; Q8; Q31; Q20
MW_P015	39	A	A	Q32; Q12; Q26; Q8; Q31; Q20; Q47; Q29; Q49; Q21; Q9; Q7; Q19; Q14; Q53; ...

The changes in the prediction probabilities for sequences $AS1$ and $AS2$ to be classified as their original category during patient response collection and replacement iterations are shown in Figure 5.4. The initial answer sequences $AS1$ and $AS2$ guarantee the maximum probability of being diagnosed as $D1$ and $D2$, respectively, during the dynamic questioning process. As the iterations continue, the diagnosis probability for $AS1$ stabilizes, while the likelihood for $AS2$ reduces drastically to below 50%. As a result, the system's classification result for $AS2$ becomes biased towards the opposite diagnostic category ($D1$), leading to termination of the dynamic adaptive questioning process and a final diagnosis of $D1$.

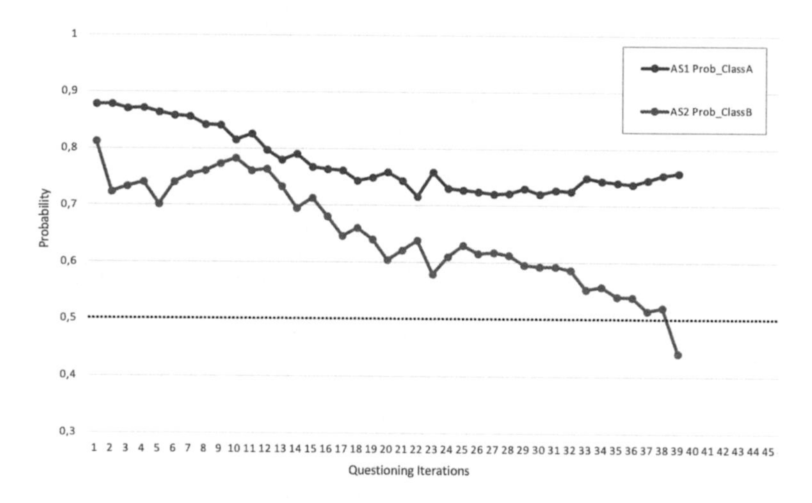

Figure 5.4: Analyzing Changes in System-Predicted Probabilities Based on Patient Response Behavior to Adaptive Question Sequences

In evaluating the performance of the dynamically formed questionnaires, the classification results of the **354** reconstructed answer sequences were compared with the doctor-confirmed diagnoses. The experimental validation results demonstrated that, compared to the static questionnaire, the dynamic adaptive questionnaire can flexibly reduce the questionnaire complexity while efficiently collecting representative factors and maintaining stable diagnostic accuracy in the RD screening process. The utilization of the patient-oriented dynamic channelized questioning method improves the diagnostic support system's capability to derive a correct prediction to **83.33%**. In contrast, the diagnostic accuracy of the system with the static questionnaire that contains **53** questions achieves **81.30%**. Whereas the questionnaire optimization implemented by exhaustive search only enhanced the accuracy of system diagnosis to **82.70%** with a fixed number of **41** questions. The dynamic adaptive questionnaire shrinks the calculation requirements for high-dimensional data and expands the interpretability for the associated symptoms of specific diseases. Facilitated by the adaptive disease screening system, the diagnosis process for early-stage rare disorders will no longer be disturbed by the cognitive biases of physicians, and the customization procedure reduces the time and effort cost of the subjects [113]. Moreover, the data storage construction (see Appendix B) enables the accumulation and extraction of respondent information and personalized settings, which can be invoked anywhere when necessary.

5.4 Discussion of Dynamic Diagnosing Method

This chapter aimed to conceptualize the data-driven method of reducing the complexity of the questionnaire on the premise of ensuring the accuracy of the system diagnosis in ascertaining rare disorders. The dynamic adaptive questioning technology has been successfully used in marketing research, user preference research, enterprise research, and social surveys [2, 38]. However, the application in RD screening is relatively conservative, and the support framework in combination with rare disorder characteristic patterns is not commonly available. For this reason, this study focuses on exploring the applicability of the self-adaptive questioning technique in the area of orphan disease identification.

Starting with the critical issues that facilitate the realization of the adaptive questioning process and the real-time input factor analysis, this chapter first described a concept for exploring the interdependencies between the input and target variables and the association network among different symptom-related questions. The correlation analysis of these variables helps discuss the questionnaire composition and arrange question orders. In the second stage, the brute-force method was proposed to statically reduce the number of questions to observe the performance of the ML-based diagnostic assistance, using a subset of RD data set to conduct experiments. Finally, for the specific disease identification task, a dynamic information acquisition method was introduced to obtain the most representative question responses required by the classification model with minimal questions. Based on exploratory data analysis, the Shannon entropy was utilized as the information measure in the question alternation section to determine the exposure rates of varied questions.

Meanwhile, the support technologies for the storage of questionnaires, the user interface for dynamic question presentation, as well as the analysis and preservation of the real-time computational results were discussed.

As a response regarding ***RQ*3**, although applying the dynamic question selection concept slightly enhances the diagnosis accuracy, reducing the number of questions while maintaining a reliable diagnostic result leads to a more centralized RD symptoms interpretation. One of the key findings of this experiment is that the proposed approach can handle noise and irrelevant variables better than a static question set. The dynamic adaptive conjoint questionnaire can reasonably adjust the subsequent questions' release priorities according to the assembled patient's symptom characteristics. For some response mistakes, it is also possible to further examine the swaying questions by dynamically retrieving relevant questions (for example, another question asking about the same symptom attribute). Understanding the purpose of questions during qualitative analysis can assess sequential questionnaires' potential biases or prejudice. The oriented questioning in the customized questionnaire attaches importance to specific symptoms of individuals. It facilitates efficient information collection and addresses a high consistency between selected questions and the particular patient, which supports experts focusing on the actual relevance of the patient's physical conditions regarding the perception of an orphan disease. The development of a dynamic adaptive questionnaire focuses on progressively analyzing whether the individual's responses to the previously released questions meet the evaluation criteria of a specific orphan disease. Compared with the conventional questionnaires that use predetermined static question combinations, the adaptive questioning mechanism efficiently reduces the questionnaire complexity. The individually organized questionnaire size saves patients' time and effort. Fundamentally, the performance of an ML-based prediction model invariably depends on the quality of the data used for model training. The variable association and pattern recognition derived from empirical data have significant influence over the application of the dynamic questionnaire.

This research shows that the dynamic adaptive questionnaire has more benefits than a survey with static question sets. The feedback from medical practitioners confirmed that the self-adaptive mechanism for RD diagnostics is helpful in the rapid screening stage. It can help scientists, researchers, and MDs reveal their patients' symptoms through identified variable correlation, enhance the questionnaire structure, and improve the questionnaire-based prediction model. However, medical experts do not recommend omitting some of these specifically designed questions while making a precise diagnosis for a specific orphan disease. It is preferred to spend more time completing the symptom-related questionnaire to grasp comprehensive information about the patient to prevent misdiagnosis. Medical experts from MHH recommend further research collaborations with diversified disease research projects to obtain more objective and comparable assessments. At the same time, the stability and general applicability of the dynamic adaptive questionnaire can be improved through more extensive practical exercises.

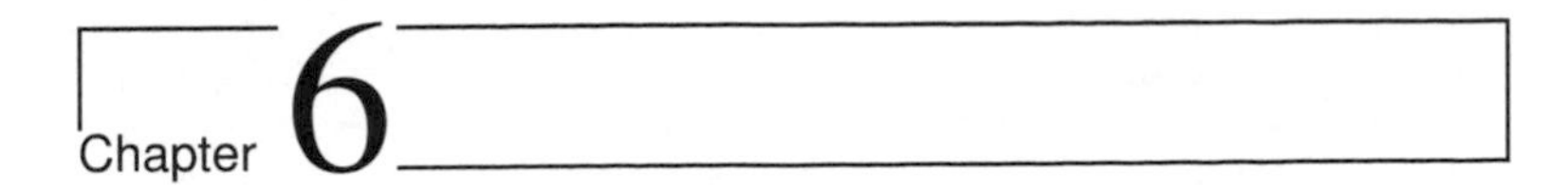

Discussion and Conclusion

With the continuous improvement of medical informatization and the impact of the digital transformation, accumulating statistically significant clinical information provide research resources to make scarce study topics accessible. As analytical technologies, data mining and ML algorithms support medical researchers in capturing critical information and discovering feasible diagnostic rules from clinical or laboratory data records. An automated disease prediction approach can promote the ability of medical systems in recognizing abnormal conditions. In the field of RD diagnosis, data-driven inference methods can assist researchers in systematically analyzing the implicit information underlying the clinical data, understanding the relationships between symptom indicators corresponding to a specific disease, detecting subtle abnormalities, and diagnosing potential patients.

This doctoral work introduced a computer-aided diagnostic support system centered on questionnaire data analysis and pattern recognition for improving the awareness of RD and diagnosing future patients. The system covers four concepts: a questionnaire topic-related response collection interface involving an automatic data cleansing process, an ML-based disease prediction model that combines four classifiers, an interactive data visualization module, and a responsive dynamic questioning mechanism combined with a real-time disease identification process. Considering the previously defined three research questions (see Section 1.2.1), this chapter provides a concluding discussion of this doctoral research by answering three research questions, respectively, together with the contribution of this research for practical RD diagnoses, as well as its limitations and prospects.

6.1 Discussion of the Research Findings

This research is motivated by analyzing RD questionnaire data through efficient ML algorithms to obtain diagnostic directions and propose a prospective RD screening assessment. Given the ambition of diagnosing future patients, the first research question addressed analytically investigating and determining the correlation between symptom-based questionnaire data and a specific RD. Learning from experts' experiences and disease features in particular domains, the decision-making support based on data mining could provide reliable perception of collected data. In this diagnostic support system, the conceptualized and practically realized classification model furnishes an integrated analysis module that utilizes rarefied disease resources and MDs' diagnostic results to predict future patients. By aggregating the respective advantages of different classifiers, the system summarizes the symptom pattern of each diagnosis category based on valid questionnaire data related to the clinic symptoms of RD patients and the diagnosis labels. The ML-based prediction model was assembled to comprehensively evaluate the probabilities of a questionnaire sequence belonging to each diagnosis category and select the most persuasive result. The application of the isotonic regression technique helps eliminate outliers and provides a calibration function for the classifier's predicted scores (see Chapter 3).

Answer to Research Question 1

The concept of the classifier fusion method utilizes an ensemble of ML algorithms to determine and refine the association between manifested symptoms and diagnostic outcomes from previously diagnosed patients' questionnaire data. The inductive and reasoning ability of different but complementary data mining techniques promote the identification of RD characteristics and diagnostic determinants. Essential diagnosis rules can be explored from the collected questionnaire data in order to assist the prediction model in identifying new RD patients. However, the accuracy of this prediction model was revealed to be highly dependent on the validity and quantity of the analytical data set that is labeled. Consequently, the proposed computer-aided decision support method might not be generalizable for determining undefined diseases or RDs with scarce research samples or unlabeled questionnaire records.

Identifying precise clinical symptoms and evaluation indicators are the key to making correct medical decisions. The LOOCV method was applied to three data sets (RD, PID, and BC) to evaluate the performance of the disease prediction model. The validation results showed that compared to involved individual classifiers, the calibrated classifier ensemble approach has robust disease identification ability and improved average diagnostic accuracy (see Section 4.2). During the implementation of this ML-based disease prediction concept in several applications (introduced in Appendix D), the received feedback corroborated that the supportive classification model can help medical practitioners identify patients with RDs, and provide an effective support for improving the accuracy of RD diagnosis in medical research and clinical practice.

While analyzing the relevance of disease symptoms and supporting RD prediction through ML methods is a vital aspect of this investigation, it is equally necessary for researchers and medical practitioners to comprehend the reasons for classification models making decisions. A highly complex data-driven learning method comes with unparalleled predictive power in identifying relationships (e.g., disease patterns), but it also makes the analysis logic difficult to understand. Due to this consideration, the second research question of this thesis thematized how to observe answer pattern differences across various RDs and how to identify essential factors in diagnosing RDs. To solve this research question, a visualization module was accomplished to assist medical practitioners in visually depicting the RD patterns and complex symptom dependencies, as well as the foundation for the classification model to make decisions (see Chapter 4).

Answer to Research Question 2

The visual infographic concept incorporated with the disease classification model emphasizes the characteristics of RDs by presenting disease symptom patterns. The Interactive information visualization function supports MDs to compare similar or mutually exclusive symptom characteristics of different diseases, share the understanding of complex associations between symptoms, and address comprehending deviations by combining domain knowledge. In addition, it helps researchers observe the response patterns of investigated individuals and discover abnormalities that deviate from the regular answer patterns. However, disease patterns or implicit information emerge based on a certain amount of data. If the sample size is too small, the data visualization results would be revealed to fail to present patterns, deviate from core information, and bias inductive ratiocination about the actual situation.

Throughout workshops and group discussions during the research period, the value of the data visualization module was recognized by MHH medical experts and practitioners, as well as the application in multiple medical research projects (see Appendix D). With a particular focus on non-data analysts, the visualization concept improves the cognitive interpretability of the data analysis results and helps MDs identify the difference between the experimental and control groups. The mapping function (see Section 4.1.2) brings benefits in reviewing the consistency of patient symptom indicators regarding the disease manifestation patterns. Moreover, the visual presentation of the single variable's influence experiment (see Section 4.2.2) and dimensional reduction results help physicians understand the significance of specific symptoms and conduct disease-matching investigations for suspicious question responses. Overall, the concept of data visualization achieves the purpose of helping medical practitioners deepen their understanding of and confidence in the prediction results of the ML-based diagnostic support system.

Characteristic patterns of different RDs summarized from previously diagnosed patients can be utilized to support the information acquisition and early-stage disease screening for future patients. Intrinsic associations between RDs and their symptoms can be found through the analysis of symptom-related questionnaire data. The application of these associations with the reference of theoretical knowledge can facilitate targeted and individualized information collection, thereby achieving channelized disease predictions. Consequently, the third research question of this thesis concerned the adaptive adjustment of the questionnaire constitution during the dynamic classification process while promoting diagnostic accuracy. A questioning mechanism based on real-time discrimination was proposed to centralize specific RD-related questions according to the gradual investigation of an individual's health condition (see Chapter 5).

Answer to Research Question 3

The dynamic questioning mechanism utilizes the correlations between disease symptoms and the diagnostic results to establish an association inference method for determining RDs. By dynamically interacting with respondents in context, the mechanism deploys the questions that are the most beneficial to diagnosing a specific disease, collects relevant information in a targeted manner, and provides the individual respondent with a unique and personalized questionnaire. The logical associations make the content of a questionnaire more reasonable and channelized for RD screening. Gathering RD-centered information enabled the exclusion of symptoms data with less relevance to the investigated RD, enhance survey response times, and speed up the decision-making process. However, the dynamic adaptive questioning and diagnosis mechanism is built upon qualified models obtained by iteratively learning the training data with precise disease labels. The mechanism does not apply to RDs with undocumented disease patterns or limited symptom indicators.

In the field of medical services, establishing a comprehensive diagnostic and predictive auxiliary system brings clinical value and social benefits to the investigation and diagnosis of orphan diseases. This doctoral research intends to assist medical practitioners in making correct decisions, optimizing the allocation of medical resources, and reducing possible misdiagnosis and missed diagnoses. The advantage of this research lies in the self-learning classifier ensemble model embedded in the diagnosis support system, which can gradually improve its prediction performance in the process of continuously accumulating empirical data. The feedback from MHH's medical experts revealed that the ML-based system not only meets the rigorous requirements of experimental research but also has the characteristics of eliminating tedious manual data operation processes and strong practicability.

In the long-term perspective, the concept of ML-based decision support is expected to increase the general awareness of RDs and the corresponding proportion of attention in the disease screening process. The validity of the established computer-aided diagnostic support concept has been demonstrated through the performance evaluation during experimental studies of RD, PID, and neuromuscular diseases data sets from MHH questionnaire data collection and BC data set from the open access UCI machine learning repository [70]. The research findings indicate that the proposed ML-based classification model with the coverage of the classifier fusion approach is universally applicable to any questionnaire with the same data structure. The scalability and feasibility of the adaptive questionnaire concept are able to find possible applications in the fields of marketing, finance, customer service, or insurance, but require complementary evaluations with data from respective fields.

6.2 Research Limitations

The research findings suggest that statistical analysis and ML algorithms facilitate the development of new approaches to safeguard the screening and diagnosis of orphan diseases. The proposed classifier fusion method can extract patterns from observed RD questionnaire data and seek regularity in RD diagnoses to help medical experts and researchers focus on critical information. However, there is no universally applicable solution that solves all RD problems [275]. Some limitations and deficiencies should be considered to pave the way for future research.

6.2.1 Limitations of the Data-driven Diagnostic System

A significant obstacle to studying RDs is the lack of research objects. In the research of ML-based diagnostic support systems, adequate data with confirmed diagnosis affects analytical precision and general applicability of disease symptom patterns. It is challenging to seek contextual connections and prove the correlation significance of the research content using scarce study resources.

Since the system does not have the cognitive and self-correcting capabilities for similar disease patterns but has different class labels, caution is required when merging data sets from multiple sources. For instance, the diagnostic results of each data category must be validated to avoid mislabeled diseases. However, RD research data are usually collected from a few sources, such as hospitals or research centers that specifically established RD research projects and local self-help groups. Patient's medical records with their diagnoses are often interior disclosure in specialized medical institutions. Due to the scarcity of RD cases, it is difficult for an independent organization to possess enough data resources for detailed and comprehensive disease analysis.

In response to this limitation, it is imperative to establish trust and acceptance for a centralized data warehouse, with the target of promoting RD questionnaires and attracting more experts in the field of rare disorders. Realizing RD patients' data sharing among multiple cooperative organizations and implementing interdisciplinary disease analyses would maximize the potential benefits of the RD symptom-based questionnaire data.

Collecting data anonymously from different sources and combined analysis results will improve the accuracy of ML-based diagnoses, and reduce the risk of false diagnoses or treatment delays. The effective integration and utilization of RD data from clinical trials across different locations have significant benefits for individual physicians, personalized patient care, rehabilitation centers, health insurance institutions, and medical researchers. However, the utility of the questionnaire data requires the authorization of the relevant RD patients. The feasibility and applicability of the data storage structure mentioned in Appendix B need to be further improved and verified.

6.2.2 Limitations of the RD Symptoms Coverage

Another prevalent problem is acquiring complete symptom-related RD data. The interpretability of RDs might be limited by geography, the professional background and empathy competency of experts involved in questionnaire design, the number of accurately diagnosed participants, and the patient's ability to comprehend the answers. It means that the symptoms of a specific RD must be recognized holistically, and the established questionnaire must elicit the symptom properties understandably to gather patient information correctly.

In addition, the final diagnosis of the responded data must be confirmed by the medical expert. Accurate diagnosis of RDs is difficult because early symptoms of the disease are sometimes hard to recognize or similar to common diseases. And not every RD can be screened by designing a specific symptom-based questionnaire. For example, RDs without significant symptoms or RDs that can be only identified by genetic testing would not be applicable to the computer-aided disease screening and diagnostic support system.

In this study, all symptom-based questionnaire data and corresponding diagnoses were confirmed on a clinical basis by medical experts. The clinical assessments of different disease groups and the collection of questionnaire data were authentic, comprehensive, and valid. However, the received feedback and recommendations during the application of this computer-aided diagnosis support system were only subjectively and experience-based diagnoses from medical experts with research backgrounds at MHH.

Furthermore, the data collected in this study were only representative of the local disease population at the time that this research program started. Due to different regions and genetic differences between various races, changes in the environment, and time, it remains to be verified whether there are differences in the specific symptoms of the same RD in different populations or generations. For example, the research results of orphan disease symptom patterns for the European region (mainly in Germany) are not necessarily applicable to Asia or Northern America. The generalization ability of the questionnaire-based diagnostic support to other RDs still requires further verification through the accumulation of questionnaire data from different categories.

6.2.3 Limitations of the Classification Model

Well performing machine learning models on finite RD data sets do not imply general applicability. Correct decision-making in the medical area is crucial and often involves complex interactions with the environment and humanistic background [205, 239, 260]. Although data records of patients can cover as many scenarios as possible, the potential events are almost infinite. The diagnostic complexity for RDs can be sophisticated and beyond the cognizance of non-domain experts.

A comprehensive ML-based diagnostic support system should exceed the diversified computational models and be able to detect linear or nonlinear relationships with assurance of accuracy. However, the appropriateness of different ML algorithm needs to be inspected based on data analysts' expertise, experience, and mathematical methods, which cause an extensive need of human effort, data, and financial resources.

In order to obtain a reliable prediction model for targeted RD data sets, the deployment and optimization process of the tailored classifier fusion method undergoes several iterations of model selection and evaluation concerning the characteristics of collected questionnaire data and the research target (see Chapter 1.2). For the selected components of the classifier fusion method, various ML algorithms were examined according to the algorithm's computation speed, accuracy, and complexity. For example, artificial neural networks (ANNs) were discarded in the experimental research because of their computational costs and time under the premise of the same accuracy level.

Four ML models were identified for the core components of the diagnostic support system to achieve an optimal classification solution with current research targets. However, accumulating and analyzing unknown symptom correlations and disease patterns of various RD questionnaire data and converting them into ready-to-use learning models is a long-term and arduous task. In this study, by experimenting and comparing the performance of different ML algorithms on the collected RD datasets, and evaluating different classifier combination models, a systematic understanding of the classifiers' applicability and potentiality is realized. But according to the diversity of disease types and data patterns, a plentiful of optional ML methods that can be flexibly combined in analyzing linear or non-linear correlations should be introduced to meet the requirement of RD questionnaire data analysis and accurate prediction.

6.2.4 Limitations of the Visualization Model

For medical experts, it is particularly important to precisely interpret the analysis results of the diagnostic support system with their domain knowledge. Since the output results of learning models are not always interpretable by humans, abstract results require additional visual concepts for the explanation. The visual representation of the data empowers medical specialists to better understand the analysis results presented by the system, and observe correlations between symptom variables and predictions. However, data visualization occurs on the human-machine interaction, not all ML results can be effectively displayed through data visualization. The visual form only shows the overall trend and schemata of the analyzed data, detailed changes would not be noticed. Due to limited medical collaborators in the research environment, the visualization concept was realized and implemented according to the responses from MHH medical experts. The impact and usefulness of visualization methods still need to undergo additional assessments involving multitudinous medical experts.

Consequently, a limitation of data visualization is regardless its simplicity and self-explanatory characteristics, the information interpretation depends on the individual perceptions of its audience. For instance, the acquirer of the data may only take notice of the considerable parts or central information, and different audience groups may interpret the graphic presentation in individual ways depending on their domain knowledge. Less attention or misfocusing on the data may lead to biased results or overlook core messages, which brings about segmented or biased interpretation results. Providing professional annotation for the visualization of RD data will improve the readability and understandability of medical researchers and analysts in non-professional fields.

6.3 Research Prospects and Future Work

Advanced improvements remain in diagnosing RDs, such as balancing the diagnostic capabilities across regions, and overcoming the shortage of open-access research findings focusing on the mechanism and genetic characteristics of the disease [9]. The effective utilization of RD patients' data is a systematic elicitation process including data processing and integration, and in some circumstances complex statistical analysis and machine learning. In the end, the interpretation of analysis results presents the final step during the data-driving diagnosis routine. Patient authorization, participation, and contribution are crucial for RD studies. The patients' feelings and feedback should be placed at the center of disease research to fulfill the understanding of the RD factors that most influence patients' health and find an effective solution based on learning from the past. In order to maximize the value of questionnaire data and provide better medical services, the following discusses the research directions and development prospects that are directly related to the future research of the ML-based diagnosis support system.

6.3.1 Expand the Coverage of RD Questionnaire Investigation

Due to the particularity of RDs, the standardization of clinical information has been a major problem for research data collection. In RD research, insufficient relevant information and knowledge popularization make it difficult for doctors to diagnose patients correctly. Since the patient's physical condition is personal privacy, different hospitals have their diagnosis and treatment recording systems and are kept confidential from other organizations. The communication impediments for RD information led to difficulties in comparing and cross-referencing disease indicators across different patients. Standardizing and anonymizing the collection and integration process of RD patients' data and rationalizing its utilization through advanced analytical techniques is the primary aspiration of ML-based RD prediction and pattern recognition research.

As the follow-up measure to strengthen the consummation of RD questionnaire databases, the questionnaire system shall be extended into different languages and popularized in cross-border medical institutions for collecting clinical manifestations of different races in other regions. The content of the questionnaire and the corresponding online operation environment were translated into German, English, Chinese, and French and approved by professionals. In the future, the questionnaire content will be spread to more countries and increase RD research cooperation intensities with the consideration of local restrictions and privacy rules. The advantage of joint data warehouses is to help explore the geographical differences under the same RDs, monitor and evaluate the disease distribution, and analyze symptom differences between different races, especially those in the specialized field of unconventional disease. An adequate data foundation can guarantee the value of discovered knowledge and the significance of research achievements propagation.

In support of data analysis and ML techniques, the computer-aided diagnostic system concerns the extensive collection of RD cases will effectively realize the digital archiving and communication of RD symptoms and clinical interview transcripts. Supported by diagnosed patients as references, combined with the knowledge and achievements of domain experts, the early diagnosis of future patients can be realized, and the appropriate treatment method can be formulated according to the patient's symptoms. The achievement of precision diagnosis can additionally improve the quality of clinical diagnosis and treatment. Simultaneously, systematically organizing clinical information and standardizing follow-up records are expected to benefit MDs and scientific researchers exploring RD pathogenesis and conducting prognostic studies. Sharing RD diagnostic information helps facilitate the development of telemedicine diagnostic systems and information exchange.

6.3.2 Automated Selection of Diversified ML Methods

Considering scientific analysis and data-driven learning models in medical institutions and the healthcare industries, the realization of a general-purpose, automated, and flexible ML system (AutoML) is a promising direction for future research that targets diagnosing various RDs. The continuous studies might focus on systematically developing a robust and compelling learning approach that automatically trains, evaluates, improves, and deploys ML models based on targeted data. Unlike the traditional ML process that requests manually identify problems and test various algorithms to determine the fitting one, AutoML aims to automate repetitive tasks in the ML pipeline, such as automated feature extraction, model selection, hyperparameter optimization, integrated learning, and model validation. So that the costs of time and human effort can be reduced while maintaining the model performance. Utilizing the model's capability under different data distributions, AutoML continuously evaluates the interval of the optimal solution and resamples that interval to detect the most suitable learning model for inferences, analogy, and induction, thereby reducing the risk of human error and improving the model efficiency. Under certain conditions where a single learning technique or even an integrated learning method is insufficient, the automatic assembly mechanism of multiple ML algorithms can optimize the effect of collaborative work. By encapsulating ML algorithms, the system would be able to automatically establish an appropriate ML model based on provided data. The flexibility of learning architecture redesign, autonomous hyperparameter adjustment, and the consolidation of the mining depth and intensity will bring a more qualified learning model. While the manual selection and ensemble of classifiers based on their performance will be replaced by AutoML, to further reduce costs by utilizing intelligent methods.

In addition to technical innovation, straightforward concepts will be easier to grasp and practice for relevant personnel of different backgrounds. The effective combination of summative qualitative and quantitative research will facilitate the interpretation of disease patterns and symptom relationships while studying diverse aspects of a phenomenon. Further research extension should focus on integrating different statistical inference and multidimensional visualization methods to generate graphics with explanatory texts that help the public unscramble a classification task and understand system decisions in a non-professional language. It is prospected to promote the development of artificial intelligence and predigest the deployment of task-based learning models. It is believed that the generalization of AutoML-based diagnostic support will demonstrate greater potency in the future.

6.3.3 Interconnection of Professional Knowledge on RD cases

In a data-driven age, enterprises and organizations have an increasing demand for aggregating individual opinions to make decisions based on statistical significance. A secured communication system connects computers distributed in different locations to form a interconnected structure capable of information interaction within a certain range. Many paramedical applications have added feedback and rating sections or discussion modules to accommodate extensive information collection and recommendation adoption. For example, the satisfaction feedback module in the medical consultation service can help to evaluate a physician's comprehensive quality from the patient's perspective and provide references for subsequent patients in doctor selections.

In responding to the challenge of RD diagnoses, embedding multimedia technology into the diagnostic support system promote collaboration between research institutes and clinical medicine. Establishing an interconnection platform between medical institutes and scientific research organizations is essential to enable the latest information communication and discussion (e.g., RD cases, research findings, diagnostic procedures, and professional opinions). With the support of an intercommunication platform in the medical diagnostic environment, connected medical and health departments will be able to promote the development of telemedicine services to optimize the allocation of medical resources and help RD patients seeking medical advice. The contribution of RD patients' clinical data and expert opinion synthesis are urgent needs to formulate treatment plans for future individuals. On the basis of RD characteristics visualization, appending case analysis and discussion module will help physicians accumulate clinical experience and promote awareness of expertise. A platform that shares information without geographical restrictions will accelerate RD research progress.

Furthermore, with the gradual accumulation of RD knowledge, the envisioned goal is to establish an up-to-date encyclopedia of RDs and associated orphan drugs, incorporating a directory of specialists and the latest research results. With the combination of worldwide latest research achievements, it is believed that a patient-centered innovation ecosystem will be gradually established with the objective of explicit, realistic, and efficient RD diagnostic. On the path of pursuing knowledge, the interpretation of RDs needs to be continuously improved along with the development of bioinformatics, statistical analysis, and resource sharing methods. The research directions need to be monitored and adjusted based on new data and findings, knowledge popularization, and new drug developments in order to improve patients' diagnoses and treatments comprehensively. Future diagnostic systems will have more mature adaptability and expressiveness, contributing to the development of RD clinical research networks and patient-centered services.

6.4 Summary of the Research

Orphan diseases — due to their complex symptoms, scattered and sparse patient cases, lack of domain knowledge, and imperfect screening procedures — are difficult to diagnose. Misdiagnoses and missed diagnoses result in patients visiting different doctors seeking for accurate answer, endure inappropriate diagnosis and treatment schemes, and high medical costs. In search of a solution that could improve the identification abilities for RDs, many countries have consecutively set up substantial collaboration of researchers, clinicians, and patients to benefit the exploration of RD etiologies [9].

With the goal of utilizing diagnosed patients' symptoms as an integrated analysis foundation to assist medical practitioners to improve RD diagnostic accuracy, a conceptual ML-based diagnostic support system focusing on RD symptom analysis and pattern recognition was proposed in this doctoral work. In collaboration with MHH medical experts, the collection of RD symptom information was realized in the form of questionnaire surveys. Combined with medical experts' domain knowledge and in-depth interviews with diagnosed patients, several symptom-oriented questionnaires targeted at different categories of RDs were created. To ensure the authenticity and validity of the questionnaire data, the data collection is concentrated in specialist clinics, RD centers, and self-help groups within Germany.

To improve the accuracy of diagnosis, deepen the understanding of RD characteristics, and accelerate the diagnosis procedure, this research combines statistical analysis methods and ML algorithms to comprehensively investigate RD questionnaire data. The principal components of the computer-aided diagnostic support system involve three modules: a supervised classifier fusion approach for pattern recognition and RD diagnosis assistance, which involves four ML algorithms (SVM, LDA, RF, and LR), an interactive data visualization module to facilitate the interpretation of RD characteristics and the foundation of system predictions; and the applicability exploration of the dynamic adaptive questioning mechanism focuses on customized symptom information acquisition on specific RD identification tasks.

The quantitative results regarding the RD data sets involved in this research indicated that the calibrated classification models yield productive computation considering the prediction accuracy according to each classifier's preponderance. To explore the necessity of RD diagnostic support and gain consensus among medical practitioners, the qualitative evaluation is built on interviews, group discussions, and workshops with MHH experts. The result reflected that physicians expressed positive feedback on the research findings. Especially the visual graphics of data analysis combined with system learning results provides a knowledge foundation for understanding the diagnosis inferences and supports the interpretation of RD characteristics from an overall perspective. The dynamic questioning mechanism reasonably adjusts released questions oriented on the physical conditions of the questionnaire participant to handle irrelevant variables, therefore promoting effective information collection while reducing time costs during questioning sessions.

The research findings revealed that detecting common phenomena among individuals with the same category of RDs can assist in optimizing the process of RD diagnostics. The unprecedented visibility and insight brought by establishing synthetic disease-related data analysis and visualization methods repository would assist medical practitioners in making more effective and reasonable diagnostics, and further consummate the development of medical systems and patients' prognosis. The development direction of AutoML along with the promotion of disease symptom information collection will provide an intelligent and standardized RD screening process. It is believed that in the future, scientists will increase their efforts in the scientific research and dialectical analysis of RDs, so as to provide guarantees for the effective diagnosis and treatment of RD patients.

Appendices

Rare Disease Data Acquisition

The main obstacle to diagnosing RDs in a timely and effective manner is insufficient scientific cognition and unqualified information support. For medical researchers, whether their central focus is on analyzing the mechanism, etiology, disorder pedigrees, or exploration of clinical therapies on targeted RDs, the study requires adequate experimental data. According to different data acquisition methods, data sources were primarily collected from qualitative research and quantitative research. The interpretive non-numerical data derived from qualitative research describes the attributes, properties, and quality of the subject under surveillance. In the medical area, the qualitative data were mainly from clinical observations, patient interviews, and specific case analyses. Quantitative research objectively collects numerical data, analyzes the relationship between variables, and extends the results to be universally cognitive. The quantitative data were collected from a large number of questionnaire surveys and investigations.

To gather information on patients' experiences and perceived intensity during their diagnostic journey, medical experts from Medizinische Hochschule Hannover (MHH) interviewed diagnosed patients and medical experts using the Delphi method [103]. The semi-structured qualitative interview was conducted in a face-to-face environment, which provided an opportunity for researchers to evaluate phenomena and influential factors from different perspectives. Meanwhile, it allowed interviewees to raise issues that the researchers may not have considered. After repetitious investigations, consultations, and revisions, a summary of the consensus view on the commonality of RD was generated. Colaizzi's phenomenological research method [209, 237] was applied to extract RD impacts on diagnosed patients and glean insights into the common phenomenon. By aggregating interview data, questionnaires for different diseases were developed as the primary tool for collecting data references. In order to analyze how affected patients differ in their traits and to explore which symptoms were most significant to distinguishing RDs, the designed questions were closely related to patients' life quality and have a solid correlation to typical early-stage symptoms of RDs [35, 92].

As an example, Table A.1 shows eight representative questions in a questionnaire that focuses on neuromuscular disease. The questionnaire contains **53** questions covering six aspects: patients' diagnostic and treatment journey, signs of the disease, emotional control, special skills acquired due to the disease, social environment, and daily life routines. The full content of this questionnaire can be found in Appendix E.1. In the static questionnaire model, patient-oriented questions focus on symptoms related to RD and indicators of individual physical conditions. These questions follow a relatively fixed sequence and use precise language.

Table A.1: Excerpt of Questions from the MHH RD Questionnaire

1	Do you suspect for a period of time prior to your diagnosis that something was wrong with your health?
2	Do you deliberately avoid activities or tasks that make your symptoms obvious to others?
3	Is it difficult for you to describe your complaints/symptoms?
4	Do you notice any special tricks or techniques you have developed to compensate for symptom–related limitations in mastering everyday tasks?
5	Have you been submitted to plenty of investigations without conclusive results?
6	Do you frequently suffer from fever / elevated temperature?
7	Did you notice pulmonary problems (shortness of breath during exercise, severe cough, breathing pauses)?
8	Is it true that you prefer staying at home (instead of going out/clubbing) since your complaints are obvious?

Closed-ended questions are often researchers' first choice because they limit the respondent choices to a fixed set of finite alternatives that are offered. Possible answers to each question are limited to a categorical or ordinal scale [244], which can measure and evaluate non-numeric concepts, such as satisfaction, happiness, and discomfort [188, 244]. The Likert scale answer options guided respondents to choose one appropriate position in a series of approval degrees for a question and ensured a smooth statistical analysis of the respondents' feelings and experiences intensity [107, 188, 258].

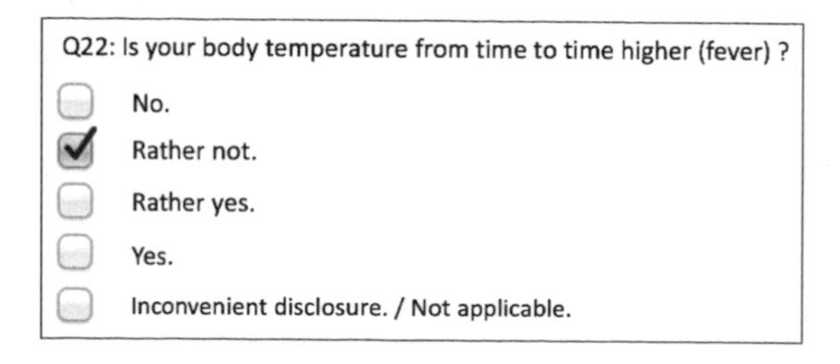

Figure A.1: Example of Answer Options from the MHH RD Questionnaire

In each question, four-level ordinal scaled answers from extreme attitude ("No") to another ("Yes") were set up to measure patients' perception and cognition of a particular symptom, as shown in Figure A.1. If the participant did not want to answer a particular question, or, in their opinion, the question does not apply to their symptoms, an answer option "Inconvenient disclosure / Not applicable" was offered. The purpose of setting up this option was to prevent respondents from inadvertently missing questions, which leads to unnecessary incomplete records.

In the data acquisition phase of this study, each participant was requested to accomplish the questionnaire independently and anonymously in order to prevent bias. By distributing the designed questionnaires to relevant medical organizations and cooperating partners in Germany, extensive data were collected from patients who suffer from different types of RDs. The composition of the questionnaire data contains quantitative information about significant symptoms of a disease, the patient's general physical constitution, and difficulties in daily activities caused by illness. Most importantly, a firm diagnosis provided by MDs was collected, and the prevalence of the participants who completed and submitted the questionnaire was verified. In practical application, the questionnaire survey can collect vast quantities of clinical information within a short time, including patient's past medical history, chief complaint, and symptoms and their development [113, 240]. It shortens the time of routine inquiry in the doctor's diagnosis process, summarizes the patient's symptoms, and organizes critical information while screening out abnormal conditions [122, 210].

Medical experts in this project streamlined and optimized the number of questions, aiming to find questions representing particular disease symptoms. The questionnaire survey has strong pertinence, efficiency, high concentration of questions, and response rate. It could quantify the research questions, and the obtained results were suitable for statistical processing and data analysis. The computer-aided diagnostic support system accomplished its expertise through a case-control study [223] and ML classification. According to the presence or absence of a particular disease, the observed subjects were divided into two groups: patients who have already been diagnosed with an RD as a disease group and non-RD individuals who have comparable symptoms as a control group. The system analyzed the correlation between symptoms and diagnosed patients by comparing the percentage differences of possible symptoms in individuals with and without an RD. The retrospective process tracing the assumed causative factors from the results can extract disease characteristics and use the answer patterns of diagnosed patients to assist doctors in predicting future patients.

Four different data sets were used in this research. Among them, **354** itemized RD questionnaire records and **126** qualified PID questionnaire samples were mainly used for training and validating the designed classification model of the computer-aided diagnostic support system. The RD data set was a subset extracted from the patients' questionnaire responses collected from the MHH research project focuses on RD diagnosis. The data set consists of **57** attributes that recorded the patients' ID, age, gender, responses to **53** questions as the measure of symptom conditions, and a diagnosis type given by an MD. The detailed data components can be found in the supporting documents of the related work [93], and the questionnaire content can be found in E.1. In order to train the model precisely, only **354** of the **1, 763** records had all of the questions answered (without null values) were selected. Within the **354** questionnaire samples, **234** records were diagnosed as rare diseases (RDs) by medical specialists, while another **120** instances labeled as other non-rare diseases (NRDs) served as the control group. Another data set was collected from a questionnaire consisting of **36** questions designed for **primary immunodeficiency disorders (PIDs)** [185]. In total, **126** qualified samples were received, including **64** PID patients who visited the *MHH - Immunological Outpatient Clinic* between **2013** and **2014** or registered at *Deutsche Selbsthilfe Angeborene Immundefekte e.V. (DSAI) - PID-Organisationen und Selbsthilfegruppen*. As a control group, **62** questionnaire records were collected from guardians of healthy children or hospitalized children due to diseases other than PID.

However, the disease spectrum varies by regions. The same disease manifests differently in different people because factors such as age, gender, physical condition, psychological diathesis, and education level impact the disease's clinical appearances. In order to reduce deviations caused by sampling from local areas, the real-world **Breast Cancer (BC)** data set, obtained from the University of Wisconsin Hospitals, Madison from Dr. William H. Wolberg [170] and retrieved from the openly accessible UCI machine learning repository [70], was employed to evaluate the system generalizability. The BC data set describes the nucleus's characteristics of benign or malignant breast masses computed from digitized images. It has **569** valid instances with **32** attributes (ID, diagnosis, 30 real-valued input features), among which **357** records are labeled as benign and **212** classified as malignant.

Additionally, since the aforementioned three data sets were designed for binary classification, a multi-class neuromuscular disease data set that contains **366** valid questionnaire records was applied to better introduce the data visualization module of the computer-aided diagnostic support system (see Chapter 4). This questionnaire involved **46** questions that focus on muscle weakness symptoms, and the received questionnaire responses were divided into **13** categories, including myotonic dystrophy and myotonia (MdMy), Morbus Pompe disease (MP), amyotrophic lateral sclerosis (ALS) and so on [92]. The concrete content of the questionnaires that focus on RDs, PIDs, and neuromuscular diseases can be found in Appendix E.

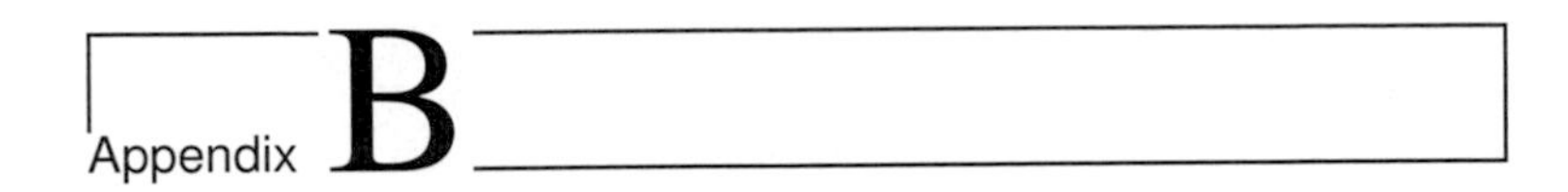

Data Storage Structure

Concerning ethical restrictions and privacy protection, patient records are protected by the supervising medical staff, department, or equalized institution that recording and using them. However, making data and analysis results shareable and open-access is crucial for optimizing medical diagnosis processes, especially in RD investigation. The value of data relies on integrating and mining potential information and patterns, transforming them into knowledge, and sharing it with others. Promoting the interoperability of RD patients' medical records can increase the research value of rare data and public awareness of RDs.

A professional data storage platform provides an environment to integrate collected data and bring convenience for researchers mining the intrinsic value of data resources thoroughly. The essential indexing services, theoretical support, and visualization products in the database provide convenience for medical experts and data analysts. Organizing questionnaire data into a database ensures the uniformity of data content, data format, and data quality provided by different sources. It reduces the duplication of labor and corresponding expenses, such as data elicitation and acquisition, and avoids the thorny problems during data disposal, thereby helping experts focus more on potential knowledge mining, developing new applications, and systematically integrating knowledge from data. The data structure of the diagnostic support system is shown in Figure B.1. Table `tbl_User` records all of the users with permission to upload data after the admin's authentication and authorization. Table `tbl_RawDataInfo` stores raw data — the uploaders are requested to manually link the raw data with the related questionnaire. The table `tbl_Questionnaire` contains the question files, targeted type of diseases, and questionnaire release information. While training the classification model, the system will automatically generates the derived information from the original data, including positive and negative predictive values, *p*-values, and the probability comparison table.

The database brings many more advantages to the modern medical system than the traditional file-based management system. First of all, the database describes

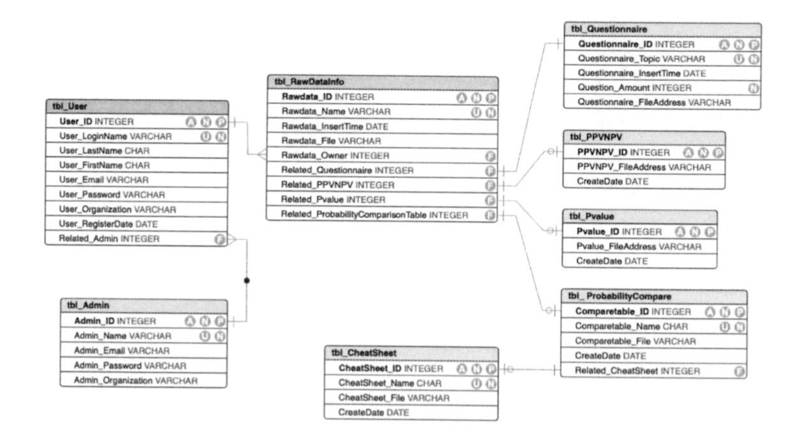

Figure B.1: Database Structure of the Computer-Aided Diagnostic Support System

and applies data from a holistic perspective and provides unambiguous relationships between different data properties [56]. The data storage is flexible and, thus, easily expandable. In addition, it can adapt to various functional requirements of different applications by extracting subsets of the overall data set. It also provides the function of periodically detecting faults and repairing them to prevent data corruption and recover from misoperations. Moreover, the database can control and manage the data centrally, and it represents and organizes data and their connections through different models. The logical structure of the data information is independent of the application, and the change in the physical data structure does not influence the logical structure. Furthermore, the database enables data sharing and interaction. All authorized users who participate in the program can easily access the data records and conveniently execute the corresponding operation and provide data sharing by using the database through various interfaces [90]. Data sharing saves storage space by significantly reducing data redundancy while avoiding incompatibilities and inconsistencies between data, guarantees the maintainability of data, and ensures data security and reliability.

Because of data collected from multiple heterogeneous sources are susceptible to noise (incorrect attribute values), missing value, inconsistencies or incomplete data, improve the data quality in the initial stage of data mining is important. Detecting data anomalies, adjusting data, and reconciling the data to be analyzed helps improve the accuracy and performance of the excavation process, thereby improving the quality of classification and prediction. In the computer-aided diagnostic support system, several techniques were applied in the data pre-processing procedure. Data cleansing helps to fill in missing values, smoothen noise data, eliminate errors, and correct inconsistencies. Data integration merges data from a plurality of sources into a consistent data storage. Data normalization eliminates the unit limit of the data and converts it into a dimensionless numerical value so that indicators with different units or magnitudes can be compared and weighted to improve the accuracy and effectiveness of mining algorithms involving distance metrics. Data reduction uses methods such as aggregation, delete redundancies, and clustering to look for features that express dependencies on the research target. It also minimizes the size of the data set while maintaining the main characteristics of the original data [109].

Figure B.2 shows the interface of data uploading and the pre-processing procedure. When an authenticated user uploads a local data set, the data source type and attributes must be selected to define the location of the data set in the database. The system will organize and save the questionnaire data by manually defining the scope of question attributes and target classification columns. Unnecessary interference attributes will be erased, which helps match multiple data sets based on unified attribute names to give direction to the backstage data integration process. By manually defining the missing value, the system will calculate the mean value of all non-missing observations under the missing value attribute to replace the missing value for interpolation.

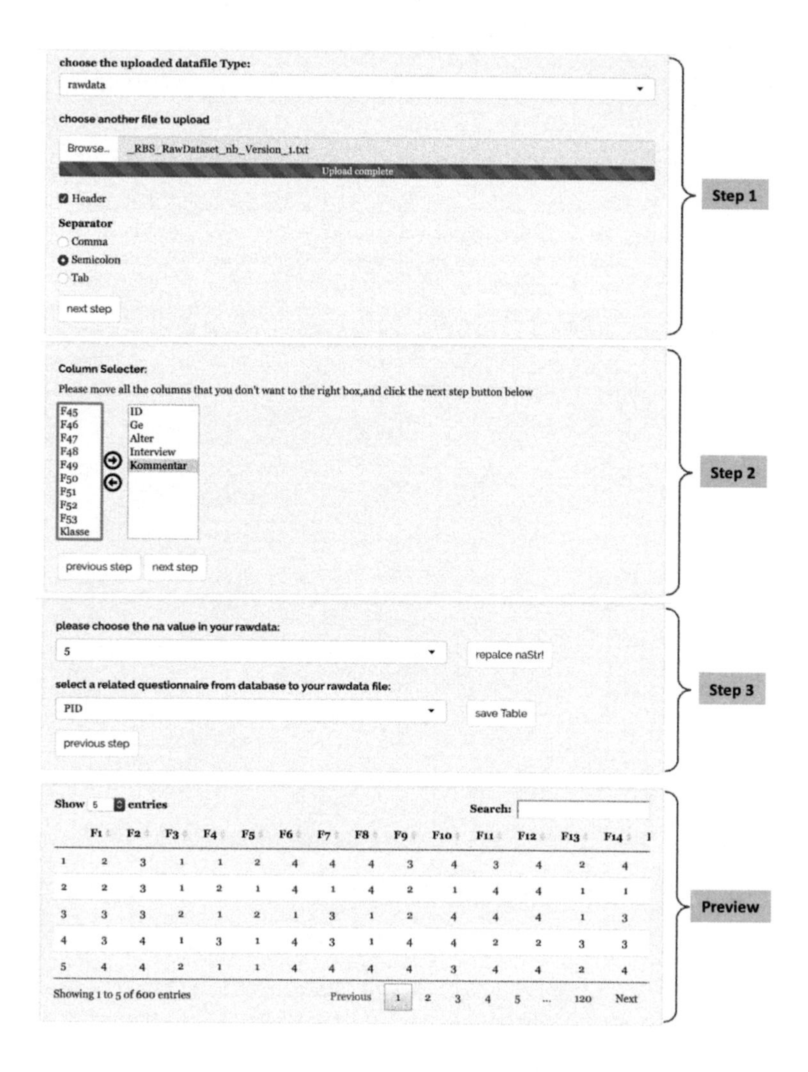

Figure B.2: The Data Uploading and Pre-Processing Interface of the Computer-Aided Diagnostic Support System

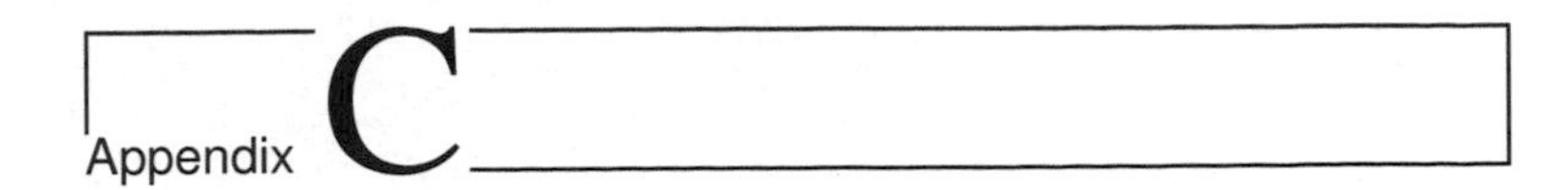

The Diagnosis–Question Correlation Ranking

Table C.1: Correlation Ranking of Individual Questions (53 in Total) with Diagnosis Class A, based on the MHH RD Dataset

Beginning of Table			
Diagnosis	**Question**	**MIC**	**Pearson**
Class A	Q19	0.12137	0.38079143
Class A	Q16	0.07503	-0.3120567
Class A	Q17	0.05597	0.27181476
Class A	Q44	0.04345	0.24193314
Class A	Q18	0.03822	0.22939944
Class A	Q38	0.03801	0.19690429
Class A	Q39	0.03798	0.22273186
Class A	Q28	0.03403	-0.16554044
Class A	Q46	0.02852	0.14722829
Class A	Q27	0.02755	-0.1943298
Class A	Q26	0.02379	-0.17861602
Class A	Q12	0.02127	0.11001853
Class A	Q45	0.01851	0.12328711
Class A	Q43	0.0184	0.15752181
Class A	Q13	0.0174	0.15346618
Class A	Q11	0.01613	0.14405255
Class A	Q4	0.01602	0.13417508
Class A	Q2	0.01564	0.14455573
Class A	Q29	0.01552	0.13711202
Class A	Q25	0.01358	0.12787
Class A	Q31	0.01154	-0.12701169

Continuation of Table C.1			
Diagnosis	**Question**	**MIC**	**Pearson**
Class A	Q1	0.01087	0.119960874
Class A	Q51	0.01069	0.045654558
Class A	Q47	0.00989	-0.11725674
Class A	Q14	0.0088	0.09838594
Class A	Q3	0.00854	0.02554033
Class A	Q5	0.0084	0.05844612
Class A	Q48	0.00828	0.0100139035
Class A	Q21	0.00787	-0.07623005
Class A	Q23	0.00735	0.09760252
Class A	Q22	0.00704	0.0108913565
Class A	Q10	0.0068	0.042824432
Class A	Q52	0.00666	-0.089086995
Class A	Q53	0.00628	-0.029677227
Class A	Q32	0.0061	0.08935488
Class A	Q49	0.00606	0.053683028
Class A	Q6	0.00591	0.07485949
Class A	Q15	0.00564	0.07525778
Class A	Q37	0.00539	0.080825716
Class A	Q9	0.00537	0.04875631
Class A	Q36	0.00499	-0.029321546
Class A	Q7	0.00436	-0.045031246
Class A	Q8	0.00368	0.05750443
Class A	Q35	0.00325	-0.018694198
Class A	Q20	0.00312	-0.035582595
Class A	Q42	0.00291	-0.04549325
Class A	Q41	0.00271	0.059617005
Class A	Q40	0.0027	-0.044365425
Class A	Q33	0.00268	0.043168023
Class A	Q24	0.00267	0.050890967
Class A	Q50	0.00156	-0.020789558
Class A	Q30	7.6E-4	-0.006811853
Class A	Q34	5.0E-4	-0.0030927234
End of Table			

Publications in the Medical Field

This appendix concentrates on collecting medical publications and scientific outputs of research projects dedicated to improving RD recognition and diagnosis technology. From **2016** to **2019**, five relevant research papers were published in prestigious medical academic journals (summarized in Table D.1). Some of these studies focus on identify rare disorders from common diseases, some put emphasis on pediatric PIDs, and some lay stress on rare neuromuscular diseases or metabolic diseases. The study results have shown that the implementation of an ML-based diagnostic support strategy facilitated the progress in pattern recognition and prospective disease prediction. These medical studies bring positive feedback to the application of the computer-aided diagnostic support system. The disease recognition and classification method are feasible and practical. Systematically analyzing patients' experiences and common phenomena promotes medical practitioners' sensitivity to essential characteristics of RDs. The data visualization provides convenience in reviewing and distinguishing RDs' clinical symptoms and improving the traceability and intelligibility of reasonable diagnosis.

Table D.1: List of Medical Research and Publications Co-authored with MHH

1	Grigull, Lorenz, et al. "Diagnostic support for selected neuromuscular diseases using answer-pattern recognition and data mining techniques: a proof of concept multicenter prospective trial." BMC medical informatics and decision making 16 (2016): 1-9.
2	Blöß, Susanne, et al. "Diagnostic needs for rare diseases and shared prediagnostic phenomena: Results of a German-wide expert Delphi survey." PLoS One 12.2 (2017): e0172532.
3	Mücke, Urs, et al. "Patient's experience in Pediatric Primary immunodeficiency Disorders: computerized classification of Questionnaires." Frontiers in immunology 8 (2017): 384.
4	Sieg, Anna-Lena, et al. "Künstliche Intelligenz zur diagnostischen Unterstützung ausgewählter seltener lysosomaler Speichererkrankungen: Ergebnisse einer Pilotstudie." Klinische Pädiatrie 231.02 (2019): 60-66.
5	Grigull, Lorenz, et al. "Common pre-diagnostic features in individuals with different rare diseases represent a key for diagnostic support with computerized pattern recognition?" PLoS One 14.10 (2019): e0222637.

D.1 Diagnostic support for selected neuromuscular diseases using answer pattern recognition and data mining techniques: A proof of concept multicenter prospective trial

ABSTRACT

Background: The correct diagnosis of neuromuscular disease is challenging. Rare diseases such as Pompe disease are easily overlooked by MDs. With the target of developing a diagnostic support application that utilizes answer patterns of patients with diagnosed neuromuscular disorders for predicting future patients. a prospective research that using statistical analysis and machine learning algorithms to analysis patients questionnaire data is conducted.
Methods: Through interviewed **16** patients diagnosed with neuromuscular disorders, MHH experts created a questionnaire containing **46** symptoms-oriented questions focused on patients' pre-diagnostic symptoms and experiences. The questionnaire was disseminated through the neurological outpatient clinic in MHH and local self-help groups in Germany. In the experimental study, seven diagnostic groups were selected for the data mining research — including myotonic dystrophy and myotonia (MdMy), Pompe disease (MP), amyotrophic lateral sclerosis (ALS), polyneuropathy (PNP), spinal muscular atrophy (SMA), other neuromuscular diseases (OND), and no neuromuscular disease (NND). During the first period, **210** questionnaire records were collected to train and validate the disease classification model that involved six different ML algorithms. In the second period designed for evaluating the model's correctness, **64** new patients' questionnaire data were collected during the one-year prospective trial.
Results: The 21-fold cross-validation results showed that among the collected **210** records, **89.5%** of the patient data were assigned with correct diagnoses. The sensitivity of the disease prediction method was **93%** – **97%** for individuals with MP, MdMy, and NND, but only **69%** for SMA patients and **81%** for ALS patients. In the second period, it was confirmed that **57** out of **64** (**89%**) diagnoses were predicted correctly by the computerized system.
Conclusion: The research findings revealed that individuals with the same diseases shared the same symptom pattern, which can be use to identify future patients. The receiver operating curve (ROC) and p-value analyses confirmed that the proposed data mining methods exhibited good results in predicting the selected neuromuscular diseases, and the fusion of different classifiers outperforms other individual classifiers in improving the diagnostic accuracy. Diagnostic support is needed for RDs due to the lack of experience in MDs and sub-specialties. However, due to the diversity of neuromuscular diseases, additional analysis must be addressed before integrating the system into routine clinical use.

Manuscript status: Published in *BMC Medical Informatics and Decision Making* 16 (BMC MED INFORM DECIS) Article number: 31.
Publication date: 08 March, 2016
Publisher name: Springer
DOI: https://doi.org/10.1186/s12911-016-0268-5

D.2 Diagnostic needs for rare diseases and shared prediagnostic phenomena: Results of a German-wide expert Delphi survey

ABSTRACT

Background: Approximately **7,000** recognized rare disease types worldwide. In particular, **4** million RD individuals in Germany. Traditionally, general practitioners (GPs) are the gatekeepers to initiate or recommend further evaluation for patients with (suspected) RD. However, complicated factors are intertwined that make diagnosing rare diseases very difficult, and it is not realistic for GPs to be able to identify all RDs. Many patients face long periods of undiagnosed, missed, or misdiagnosed conditions. To identify commonalities among patients with different rare diseases and explore the need for diagnostic support, this study conducted a two-step Delphi survey utilizing web-based technology, covering the patient population in RD centers across Germany.
Methods: Individuals from the German foundation for rare diseases (ACHSE) was invited as experts to list down the name of RDs that has special requirements for diagnostic and treatment improvement. Meanwhile, common symptoms and experiences of RD patients were recorded and analyzed. During the Delphi process, a total of **166** individuals participated in the first step of the survey, which primarily involved physicians and affected individuals or relatives, and **95** of them participated in the second round.
Results: Participants announced that **456** rare diseases or disease groups need diagnostic support. Especially Metabolic, autoimmune diseases, and three specific diseases (scleroderma, Pompe's disease, and pulmonary arterial hypertension) stated a high demand for diagnostic support. Commonalities exist among different RD patients in the early stage of diagnosis. During the survey period, a total of **373** different commonalities were collected, of which **183** commonalities were categorized as "special emotions and cognition of RD patients", and **80** quotations to the category of "endured years of visiting medical specialists and inaccurate therapy before receiving confirmed diagnoses".
Conclusion: The study results revealed that there is a high demand for the diagnostic support for patients with RDs. Common prediagnostic phenomena of RD individuals and their experiences can be utilized to analysis RD patient patterns, therefore serve as the foundation for developing diagnostic support tools.

Manuscript status: Published in *PLoS One* 12(2): e0172532.
Publication date: 24 February, 2017
Publisher name: Public Library of Science
DOI: https://doi.org/10.1371/journal.pone.0172532

D.3 Patient's Experience in Pediatric Primary Immunodeficiency Disorders: Computerized Classification of Questionnaires

Abstract

Background: Primary immunodeficiency disorders (PIDs) are a heterogeneous group that involves more than **200** RDs. Prompt diagnosis is critical to the prognosis of rare diseases. The purpose of this study was to develop a symptom-oriented diagnostic questionnaire combined with ML-based pattern recognition methods to support MDs in identifying suspicious PID patients.
Methods: Through standardized interviews with guardians of children who suffered PID. A questionnaire based on parental observations was developed using Colaizzis' content analysis framework. During the study period, the questionnaire that contains 36 symptom-related questions received 126 valid responses. Six ML algorithms — support vector machines, random forests, logistic regression, naïve Bayes, linear discriminant analysis, and k-nearest neighbor — were trained by questionnaire responses from **64** PID patients and **62** non-PID patients to establish the diagnostic support tool. A fusion method was used to combine six classifier's prediction results into one final decision.
Results: The11-fold cross-validation was applied to evaluate the performance of the classification model. The validation results showed that the classifier fusion method achieved an overall diagnostic sensitivity of **98%**. **113/126** (**90%**) of the questionnaires were classified correctly. Regarding the group of children with PID, **63** of **64** (**98%**) patients received correct diagnoses.
Conclusion: The approach in the study was proved to be suitable for identifying disease patterns and thus distinguishing PID patients and non-PID patients. The research findings provide a basis for developing diagnostic support tools for determining RD patients through questionnaires, and it revealed the diagnosed patients' experiences are considerable references for identifying new patients.

Manuscript status: Published in *Frontiers in Immunology* 8 (pp 384).
Publication date: 05 April 2017
Publisher name: Frontiers Media SA
DOI: https://doi.org/10.3389/fimmu.2017.00384

D.4 Künstliche Intelligenz zur diagnostischen Unterstützung ausgewählter seltener lysosomaler Speichererkrankungen: Ergebnisse einer Pilotstudie

Abstract

Background: To assist doctors in diagnosing rare metabolic diseases and increase diagnostic pathways, this research proposed a machine learning-based diagnostic method relied on patients' personal experiences and early symptoms.
Methods: The study established a questionnaire that focuses on selected rare metabolic diseases (Mucopolysaccharidosis (MPS), M. Fabry and M. Gaucher) by conducting **17** interviews with diagnosed individuals or their guardians. Four ML algorithms (including support vector machine, random forest, logistic regression, and linear discriminate analysis) were selected to detect suspicious rare metabolic disease patients' answer patterns in collected questionnaires. A classifier ensemble method was proposed to deliver the final diagnosis of four trained classification models.
Results: 8-fold cross-validation was applied on **56** valid questionnaire data to evaluate the performance of the classifier ensemble model. In each validation iteration, **49** questionnaires were used for model training, and **7** questionnaire records were used for testing. The validation result showed that the aforementioned classification model achieved a sensitivity value of **91%** for the diagnose group 'MPS'. As a preliminary prospective test, additional **20** questionnaire records that have not been used for the training process received **18** correct diagnoses (**90%**).
Conclusion: The research findings proved the potential of using data mining and pattern recognition techniques to diagnose future patient. Utilizing collected questionnaire data that have confirmed diagnosis labels, the classifier fusion approach showed its validity and practicability for diagnosing selected metabolic diseases.

Manuscript status: Published in *Klinische Pädiatrie* 231(2) (pp 60-66).
Publication date: January, 2019
Publisher name: Georg Thieme Verlag KG
DOI: http://dx.doi.org/10.1055/a-0816-5681

D.5 Common pre-diagnostic features in individuals with different rare diseases represent a key for diagnostic support with computerized pattern recognition?

ABSTRACT

Background: Rare diseases have a variety of clinical manifestations, and making traditional clinical diagnoses based on MDs' experience is a considerable challenge. Another challenge faced by patients in RDs diagnosis and treatment is that fewer specialists are focusing on orphan diseases than patients. Previous studies have shown that there are some common phenomena among individuals affected by (different) RD before the diagnosis is confirmed. The research aims to find commonalities between different RDs and develop ML-based diagnostic support for RDs.
Methods: The study extracted the key phenomena of RDs and created a questionnaire by conducting interviews with **20** individuals that suffered from different RDs combined with the diagnostic experience of medical professionals. The questionnaire — which contains **53** questions — was distributed to individuals that were diagnosed with RDs, common non-rare diseases (NRDs), chronic diseases (CD), and psychosomatic or somatoform disorders (PSY). The recovered questionnaire responses were analyzed through a fusion algorithm composed of four different classifiers (involving a support vector machine, random forest, logistic regression, and linear discriminate analysis) to distinguish different patient groups' response patterns.
Results: In total, **1763** questionnaire records were collected, including **758** RDs, **149** CDs, **48** PSYs, **200** NRDs, **34** healthy individuals, and **574** non-evaluable questionnaires. The 10-fold cross-validation was applied to three independent data sets to evaluate the performance of the proposed classifier fusion model. The validation results revealed that the sensitivity value of detecting RD is **88.9%**, and for NRD is **86.6%**, **87.7%** for CD, and **84.2%** for PSY.
Conclusion: The study found that although the symptoms and pathogenesis of each type of RD varied, there were substantial similarities in the experiences of RD patients along their diagnostic journey. Data mining and machine learning methods based on questionnaire data can successfully detect unique patterns in individuals affected by RD. Therefore, the research findings suggest that the proposed computer-aided diagnostic support system can be used for identifying RD.

Manuscript status: Published in *PLoS One* 14(10):e0222637.
Publication date: 10 October, 2019
Publisher name: Public Library of Science
DOI: https://doi.org/10.1371/journal.pone.0222637

Appendix E

Questionnaire Content

E.1 Rare Disease Questionnaire

URSACHENSUCHE:

	Antwortmöglichkeiten:	nein	eher nicht	eher schon	ja	weiß nicht
Q1.	Ahnen/ ahnten Sie – ggf. schon länger – dass „etwas" mit Ihnen nicht stimmt?	☐	☐	☐	☐	☐
Q2.	Trifft es zu, dass Ihre Beschwerden/ irritierenden Erscheinungen bei Ärzten/ Ihrem Arzt erst einmal ohne Reaktion blieb?	☐	☐	☐	☐	☐
Q3.	Hat man bei Ihnen auffällige Untersuchungsergebnisse (z.B. Blutwerte, EKG, Hautveränderungen, usw.) festgestellt, deren Ursache jedoch zunächst nicht weiter untersucht wurde?	☐	☐	☐	☐	☐
Q4.	Trifft es zu, dass es schwierig für Sie ist/ war, Ihre Beschwerden/ irritierenden Erscheinungen in Worte zufassen?	☐	☐	☐	☐	☐
Q5.	Haben Sie es jemals für möglich gehalten, dass die Ursache für Ihre Beschwerden/ irritierenden Erscheinungen in Ihren Lebensgewohnheiten (z.B. Ernährung, Wohnverhältnisse, Reisen, usw.) liegen könnte?	☐	☐	☐	☐	☐
Q6.	Wurden bei Ihnen viele Untersuchungen durchgeführt, die ohne Ergebnis blieben?	☐	☐	☐	☐	☐

	Antwortmöglichkeiten:	nein	eher nicht	eher schon	ja	weiß nicht
Q7.	Gingen oder gehen Sie immer wieder mit denselben Beschwerden/ irritierenden Erscheinungen zu Ärzten/ Ihrem Arzt?	☐	☐	☐	☐	☐
Q8.	Haben Sie wegen unterschiedlicher Beschwerden/ irritierender Erscheinungen verschiedene Fachärzte aufgesucht?	☐	☐	☐	☐	☐
Q9.	Erinnern Sie sich – im Zusammenhang mit Ihren Beschwerden – an eine Situation, die Sie als besonders bedrohlich erlebt haben?	☐	☐	☐	☐	☐
Q10.	Setzen Sie im Alltag spezielle Techniken (Tricks und Kniffe) ein, um Ihre Einschränkungen auszugleichen?	☐	☐	☐	☐	☐
Q11.	Vermeiden Sie bewusst Aktivitäten (Tätigkeiten), bei denen Ihre Beschwerden/ irritierenden Erscheinungen sichtbar werden?	☐	☐	☐	☐	☐
Q12.	Haben Sie sich aus eigener Initiative heraus über mögliche Ursachen Ihrer Beschwerden/ irritierenden Erscheinungen informiert?	☐	☐	☐	☐	☐
Q13.	Haben Ihre Beschwerden/ irritierenden Erscheinungen im Laufe der Zeit immer wieder andere Namen (Diagnosen) erhalten?	☐	☐	☐	☐	☐
Q14.	Haben Sie im Verlauf Ihrer Diagnosesuche von sich aus nach Ärzten (Spezialisten, Experten) gesucht?	☐	☐	☐	☐	☐
Q15.	Haben Sie schon einmal die Durchführung von Untersuchungen von Ärzten/ Ihrem Arzt gefordert?	☐	☐	☐	☐	☐
Q16.	Haben Sie den Eindruck, dass die von Ihnen geschilderten Beschwerden von Ärzten/ Ihrem Arzt ernst genommen werden/ wurden?	☐	☐	☐	☐	☐
Q17.	Trifft es zu, dass Sie Ihr Vertrauen in Ärzte/Ihren Arzt verloren haben?	☐	☐	☐	☐	☐
Q18.	Hatten Sie schon einmal den Punkt erreicht, an dem Sie Ihre Diagnosesuche aufgegeben haben?	☐	☐	☐	☐	☐
Q19.	Trifft es zu, dass bei Ihnen eine psychische/psychosomatische Erkrankung (*z.B. aufgrund von Stress, Traumatischen Erlebnissen, usw.*) vermutet wird/wurde?	☐	☐	☐	☐	☐

	Antwortmöglichkeiten:	nein	eher nicht	eher schon	ja	weiß nicht
Q20.	Gab es ein bestimmtes Erlebnis, dass Ihnen die Zunahme (Verschlechterung) Ihrer Beschwerden besonders vor Augen führte?	☐	☐	☐	☐	☐

KRANKHEITSZEICHEN:

	Antwortmöglichkeiten:	nein	eher nicht	eher schon	ja	weiß nicht
Q21.	Haben/ hatten Sie häufig erhöhte Temperatur (Fieber)?	☐	☐	☐	☐	☐
Q22.	Leiden Sie an mehreren Beschwerden/ irritierenden Erscheinungen zugleich (*z.B. Husten* ***und*** *Gewichtsverlust / Hautausschlag* ***und*** *Leistungsknick* ***und*** *Sehverschlechterung*)?	☐	☐	☐	☐	☐
Q23.	Leiden Sie unter ständiger Müdigkeit?	☐	☐	☐	☐	☐
Q24.	Haben/ hatten Sie eine irritierende Besonderheit (*z.B. Verfärbungen der Haut, Größerwerden von Körperteilen, Zittern, Zuckungen, usw.*) an sich festgestellt?	☐	☐	☐	☐	☐
Q25.	Leiden/ litten Sie immer wieder unter starke Schmerzen?	☐	☐	☐	☐	☐
Q26.	Hat Ihre körperliche Leistungsfähigkeit (*z.B. im Sport, beim Treppensteigen, usw.*) spürbar abgenommen?	☐	☐	☐	☐	☐
Q27.	Haben Sie Atembeschwerden an sich festgestellt (*z.B. Kurzatmigkeit bei Belastung, starker Husten, Schnarchen, Atemaussetzer, usw.*)?	☐	☐	☐	☐	☐
Q28.	Trifft es zu, dass Sie einen unsicheren Gang haben (*z.B. weil Sie schwanken, stolpern, stürzen, usw.*)?	☐	☐	☐	☐	☐
Q29.	Waren Sie aufgrund Ihrer Beschwerden schon einmal voller Verzweiflung?	☐	☐	☐	☐	☐
Q30.	Trifft es zu, dass Sie von Menschen aus Ihrem Umfeld (*Familie, Bekannte, Freunde, Kollegen, usw.*) auf körperliche Auffälligkeiten angesprochen wurden?	☐	☐	☐	☐	☐

BESCHWERDEN UNTER KONTROLLE BRINGEN:

	Antwortmöglichkeiten:	nein	eher nicht	eher schon	ja	weiß nicht
Q31.	Haben Sie mit der Zeit gelernt, Ihre Beschwerden besser einzuschätzen?	☐	☐	☐	☐	☐
Q32.	Haben Sie von sich aus versucht, Ihre Beschwerden zu lindern (*z.B. durch Schmerzmittel, Salben, Wickel, Sport, besondere Ernährung, usw.*)?	☐	☐	☐	☐	☐
Q33.	Haben Sie bei Ärzten/ bei Ihrem Arzt schon einmal auf eine bestimmte Behandlung (Therapie) bestanden?	☐	☐	☐	☐	☐

BESONDERS SEIN:

	Antwortmöglichkeiten:	nein	eher nicht	eher schon	ja	weiß nicht
Q34.	Trifft es zu, dass Sie Dinge können, die Andere nicht können – oder anders herum?	☐	☐	☐	☐	☐
Q35.	Werden Ihnen Ihre Beschwerden/ irritierenden Erscheinungen erst im Vergleich mit „gesunden" Menschen deutlich?	☐	☐	☐	☐	☐
Q36.	Galten Sie als Kind/ im jugendlichen Alter als unsportlich (*waren Sie z.B. vom Schulsport befreit oder nahmen hieran ungern teil*)?	☐	☐	☐	☐	☐
Q37.	Stehen/ standen Sie aufgrund Ihrer Beschwerden/ irritierenden Erscheinungen häufig im Mittelpunkt?	☐	☐	☐	☐	☐
Q38.	Trifft es zu, dass Sie sich für sichtbare Veränderungen schämen/ schämten?	☐	☐	☐	☐	☐
Q39.	Sind Ihre Beschwerden für Andere verborgen bzw. äußerlich unsichtbar?	☐	☐	☐	☐	☐
Q40.	Haben Sie den Eindruck, dass andere Menschen auf Sie Rücksicht nehmen müssen?	☐	☐	☐	☐	☐
Q41.	Haben/ hatten Sie Ihre Beschwerden so weit wie möglich zurückgedrängt, im Bemühen, ein normales Leben zu führen?	☐	☐	☐	☐	☐
Q42.	Ist/ war die Ungewissheit über die Ursache Ihrer Beschwerden für Sie das Schlimmste?	☐	☐	☐	☐	☐
Q43.	Trifft es zu, dass Sie schon einmal selbst geglaubt haben, sich Ihre Beschwerden nur einzubilden (einzureden)?	☐	☐	☐	☐	☐

SOZIALES UMFELD:

	Antwortmöglichkeiten:	nein	eher nicht	eher schon	ja	weiß nicht
Q44.	Haben/ hatten Sie den Eindruck, dass Ihr Umfeld (*Familie, Freunde, Bekannte, Kollegen, usw.*) Ihre Beschwerden **nicht** ernst nimmt (*z.B. dass jemand sagt: „Ist doch nicht so schlimm...“*)?	☐	☐	☐	☐	☐
Q45.	Trifft es zu, dass Sie Ihrem Umfeld (*z.B. Familie, Bekannte, Freunde, Kollegen*) Informationen zu Ihren Beschwerden/ irritierenden Erscheinungen bewusst vorenthalten?	☐	☐	☐	☐	☐
Q46.	Trifft es zu, dass Sie lieber zu Hause bleiben (*z.B. weniger mit Freunden unternehmen, usw.*), seit Ihre Beschwerden/ irritierende Erscheinungen deutlich sind?	☐	☐	☐	☐	☐
Q47.	Haben Sie eine bestimmte Vertrauensperson, auf die Sie sich – im Zusammenhang mit Ihren Beschwerden/ irritierenden Erscheinungen – zu 100% verlassen können?	☐	☐	☐	☐	☐

ALLTAG:

	Antwortmöglichkeiten:	nein	eher nicht	eher schon	ja	weiß nicht
Q48.	Verzichten Sie auf bestimmte Aktivitäten, die Ihnen eigentlich Freude machen?	☐	☐	☐	☐	☐
Q49.	Haben Sie aufgrund Ihrer Beschwerden/ irritierender Erscheinungen Veränderungen in Ihrem Alltag (z.B. Tagesablauf geändert, alternative Wege gewählt) vorgenommen?	☐	☐	☐	☐	☐
Q50.	Trifft es zu, dass Sie Unternehmungen (z.B. Ausgehen, Tagesausflüge, Urlaubsreisen, usw.) sehr vorrausschauend und genau planen/ planten?	☐	☐	☐	☐	☐
Q51.	Greifen Sie auf Hilfsmittel zurück, um im Alltag besser zurecht zu kommen?	☐	☐	☐	☐	☐
Q52.	Ziehen Sie aufgrund Ihrer gesundheitlichen Situation eine berufliche Veränderung in Erwägung (bzw. haben Sie diese bereits umgesetzt)?	☐	☐	☐	☐	☐
Q53.	Trifft es zu, dass Sie – im Vergleich mit anderen Menschen – größere Anstrengungen unternehmen müssen, um Ihre Ziele zu erreichen?	☐	☐	☐	☐	☐

E.2 Primary Immunodeficiency Disorders Questionnaire

	Antwortmöglichkeiten:	nein	eher nicht	eher schon	ja	weiß nicht
Q1.	Did your child suffer from ill health constantly?	☐	☐	☐	☐	☐
Q2.	Did your child eat and drink sufficiently?	☐	☐	☐	☐	☐
Q3.	Do you think that further examinations (e.g. blood test) should have been done in order to verify your observations concerning your child's health?	☐	☐	☐	☐	☐
Q4.	Did your child's illness already appear "normal" to you?	☐	☐	☐	☐	☐
Q5.	Did your child regain full health between periods of illness?	☐	☐	☐	☐	☐
Q6.	Did your child's illness repeatedly affect the same body regions (*e.g. always the ear, always the lungs, ...*)?	☐	☐	☐	☐	☐
Q7.	Did your child look slim/frail?	☐	☐	☐	☐	☐
Q8.	Did your child suffer from diarrhoea?	☐	☐	☐	☐	☐
Q9.	Did you regularly notice unusual breathing sounds?	☐	☐	☐	☐	☐
Q10.	Did your child suffer from constant coughing?	☐	☐	☐	☐	☐
Q11.	Did your child have swollen lymph nodes?	☐	☐	☐	☐	☐
Q12.	Did your child suffer from high fever without the cause for the fever being found?	☐	☐	☐	☐	☐
Q13.	Did you notice a regular white-ish film in your child's mouth that was difficult to wipe off?	☐	☐	☐	☐	☐
Q14.	Did your child have a small head compared to other children?	☐	☐	☐	☐	☐
Q15.	Would you say your child was fit enough for sports and physical exercise?	☐	☐	☐	☐	☐
Q16.	Did antibiotics work satisfactorily when your child was treated with them?	☐	☐	☐	☐	☐

	Antwortmöglichkeiten:	nein	eher nicht	eher schon	ja	weiß nicht
Q17.	Do any known immunodeficiencies run in your family?	☐	☐	☐	☐	☐
Q18.	Are there any known cases of children's deaths in your family for which no clear cause could be found?	☐	☐	☐	☐	☐
Q19.	Would you say that exceptionally many members of your family have been diagnosed with cancer in the past?	☐	☐	☐	☐	☐
Q20.	Did you have to take your child to hospital?	☐	☐	☐	☐	☐
Q21.	Did doctors find any unusual blood values in your child (e.g. changes in blood count)?	☐	☐	☐	☐	☐
Q22.	Did your child suffer from pneumonia?	☐	☐	☐	☐	☐
Q23.	Did you have to consult a doctor due to recurring earaches?	☐	☐	☐	☐	☐
Q24.	Did your child suffer from meningitis?	☐	☐	☐	☐	☐
Q25.	Did you have to use antipyretic medication (per month)?	☐	☐	☐	☐	☐
Q26.	Did your child suffer from fevers that could not or not really be lowered by treatment with medication?	☐	☐	☐	☐	☐
Q27.	Is it true that you repeatedly switched doctors because the reasons for your child's health issues could not be found?	☐	☐	☐	☐	☐
Q28.	Is it true that your child's health issues increased consistently?	☐	☐	☐	☐	☐
Q29.	Is it true that everyday incidents (like not being dressed warmly enough) sufficed for your child to fall ill?	☐	☐	☐	☐	☐
Q30.	Is it true that your child's illnesses repeatedly showed the same or similar patterns?	☐	☐	☐	☐	☐
Q31.	Is it true that your child's infections lasted longer than the ones of other children?	☐	☐	☐	☐	☐
Q32.	Is it true that your child was treated with antibiotics regularly?	☐	☐	☐	☐	☐

	Antwortmöglichkeiten:	nein	eher nicht	eher schon	ja	weiß nicht
Q33.	Is it true that you consulted a doctor repeatedly due to similar or identical health issues?	☐	☐	☐	☐	☐
Q34.	Is it true that your child has always been a bit smaller than other children?	☐	☐	☐	☐	☐
Q35.	Is it true that the doctors could not tell what your child was suffering from?	☐	☐	☐	☐	☐
Q36.	Is it true that your child was absent from school/pre-school/kindergarten due to sickness more often than other children?	☐	☐	☐	☐	☐

E.3 Neuromuscular Disease Questionnaire

	Antwortmöglichkeiten:	nein	eher nicht	eher schon	ja	weiß nicht
Q1.	Ist es Ihnen problemlos möglich, einen schweren Gegenstand (*z.B. Getränkekarte*) vom Boden zu heben?	☐	☐	☐	☐	☐
Q2.	Fühlen Sie sich im Vergleich zu anderen Lebensphasen vermehrt schlapp und müde?	☐	☐	☐	☐	☐
Q3.	Geht es für Sie bequem, eine Sprudelflasche mit der Hand öffnen?	☐	☐	☐	☐	☐
Q4.	Waren Ihre "Leberwerte" schon einmal erhöht, ohne dass man sich dieses erklären konnte?	☐	☐	☐	☐	☐
Q5.	Stolpern Sie öfter beim Gehen, weil Sie mit einem Fu_ irgendwie hängen bleiben?	☐	☐	☐	☐	☐
Q6.	Können Sie Ihre Beine im Liegen gleichzeitig anheben und knapp über dem Boden halten?	☐	☐	☐	☐	☐
Q7.	Verspüren Sie nach körperlicher Aktivität Schmerzen?	☐	☐	☐	☐	☐
Q8.	Bereitet es Ihnen Schwierigkeiten, über kleine Pfützen springen?	☐	☐	☐	☐	☐
Q9.	Können Sie ein Gepäckstück bequem in Kopfhöhe verstauen?	☐	☐	☐	☐	☐
Q10.	Bemerken Sie manchmal Muskelzuckungen, die "von ganz allein auftreten"?	☐	☐	☐	☐	☐
Q11.	Werden Sie von anderen oft als bewegungsfaul bezeichnet?	☐	☐	☐	☐	☐
Q12.	Wenn Sie Ihr Gleichgewicht verlieren, können Sie dann nicht schnell genug gegensteuern?	☐	☐	☐	☐	☐
Q13.	Trifft es zu, dass nach mehrtägiger körperlicher Ruhe ihre Muskelkraft deutlich schlechter wird?	☐	☐	☐	☐	☐
Q14.	Begleiten Sie ständige Rückenschmerzen?	☐	☐	☐	☐	☐
Q15.	Haben Sie eine Art Überbeweglichkeit der Gelenke?	☐	☐	☐	☐	☐
Q16.	Wurde bei Ihnen eine Erhöhung der CK (*Creatinkinase, ein Muskelenzym*) festgestellt?	☐	☐	☐	☐	☐

	Antwortmöglichkeiten:	nein	eher nicht	eher schon	ja	weiß nicht
Q17.	Können/Konnten Sie ohne Probleme ein Bein im Stand seitlich abspreizen?	☐	☐	☐	☐	☐
Q18.	Trifft es auf Sie zu, dass Sie bei sportlichen Aktivitäten zwar oft "woll(t)en" aber nicht können?	☐	☐	☐	☐	☐
Q19.	Brauchen Sie nach Krankheit häufig länger, um wieder auf die Beine zu kommen?	☐	☐	☐	☐	☐
Q20.	Wenn Sie sich im Liegen auf die Seite (*bzw. vom Bauch auf den Rücken*) drehen, unterscheiden sich ihre Bewegungen von anderen?	☐	☐	☐	☐	☐
Q21.	Können Sie problemlos im Liegen den Kopf anheben und halten?	☐	☐	☐	☐	☐
Q22.	Fällt es Ihnen ausgesprochen schwer, bergauf zu gehen?	☐	☐	☐	☐	☐
Q23.	Trifft es zu, dass Sie Sitzgelegenheiten mit Armlehnen bevorzugen, weil Sie sich beim Aufstehen durch Hilfe der Arme abstoßen können?	☐	☐	☐	☐	☐
Q24.	Fallen Sie manchmal aus unerklärlichen Gründen?	☐	☐	☐	☐	☐
Q25.	Haben Sie Schwierigkeiten, aus der Hocke hoch zu kommen?	☐	☐	☐	☐	☐
Q26.	Kann man sagen, dass Sie beim Gehen Ihre Beine "nach außen werfen"?	☐	☐	☐	☐	☐
Q27.	Fällt es Ihnen schwer, "über Kopfhöhe" zu arbeiten (*z.B. Wäsche aufhängen oder Nagel einschlagen*)?	☐	☐	☐	☐	☐
Q28.	Konnten Sie schon in der Jugend bei ganz vielen Sportarten eher nicht mithalten?	☐	☐	☐	☐	☐
Q29.	Trifft es zu, dass Sie Schwierigkeiten (*z.B. undeutliche Sprache, Lispeln*) beim Sprechen bemerken?	☐	☐	☐	☐	☐
Q30.	Ist Ihnen eine deutliche Verminderung der Muskelkraft aufgefallen, z.B. der Arme oder Beine?	☐	☐	☐	☐	☐
Q31.	Wenn Sie im Sitzen die Beine überschlagen, unterstützen sie diese Bewegung meistens mit der Hand?	☐	☐	☐	☐	☐
Q32.	Können Sie Lasten vom Boden eher nur mit Schwung aufnehmen?	☐	☐	☐	☐	☐

	Antwortmöglichkeiten:	nein	eher nicht	eher schon	ja	weiß nicht
Q33.	Trifft es zu, dass es Ihnen schwerfällt, das Hemd zuzuknöpfen?	☐	☐	☐	☐	☐
Q34.	Bereitet Ihnen das einfache Hochspringen aus dem Stand Schwierigkeiten?	☐	☐	☐	☐	☐
Q35.	Haben Sie das Gefühl, dass einzelne Gliedmaßen (*z.B. Beine oder Arme*) an Masse verlieren (*kleiner werden / schrumpfen / zu schmelzen scheinen*)?	☐	☐	☐	☐	☐
Q36.	Können kleinste Unebenheiten im Boden Ihnen das sichere Gehen erschweren?	☐	☐	☐	☐	☐
Q37.	Fallen Ihnen Sportarten, die mit spontaner Kraftentfaltung einhergehen, wie Springen, Werfen, Sprintlauf eher sehr schwer?	☐	☐	☐	☐	☐
Q38.	Hören Sie öfter, dass Sie einen "typischen oder auffälligen Gang" haben?	☐	☐	☐	☐	☐
Q39.	Wenn Sie Ihre Beine hochlegen wollen, brauchen Sie dann manchmal die Unterstützung der Arme beim Anheben?	☐	☐	☐	☐	☐
Q40.	Trifft es zu, dass Sie sportlich Erlerntes kräftemäßig plötzlich nicht mehr konnten?	☐	☐	☐	☐	☐
Q41.	Können Sie eine volle Wasserflasche bequem am ausgestreckten Arm seitlich halten?	☐	☐	☐	☐	☐
Q42.	Leiden Sie unter Muskelkrämpfen?	☐	☐	☐	☐	☐
Q43.	Fällt es Ihnen eher schwer Lasten zu tragen, wie z.B. volle Einkaufstaschen?	☐	☐	☐	☐	☐
Q44.	Haben Sie abstehende Schulterblätter (*so genannte "Engelsflügel"*)?	☐	☐	☐	☐	☐
Q45.	Kann man sagen, dass Sie bei anstrengenden Tätigkeiten auf andere Muskelgruppen oder Bewegungsmuster ausweichen?	☐	☐	☐	☐	☐
Q46.	Ist Ihnen aufgefallen, dass Sie ohne erkennbare Gewichtszunahme einen dickeren Bauch bekommen haben?	☐	☐	☐	☐	☐

Bibliography

[1] AAMODT, A. ; PLAZA, E. : Case-based reasoning: Foundational issues, methodological variations, and system approaches. In: *AI communications* 7 (1994), Nr. 1, S. 39–59

[2] ABERNETHY, J. ; EVGENIOU, T. ; VERT, J.-P. : An optimization framework for adaptive questionnaire design. In: *INSEAD, Fontainebleau, France* (2004)

[3] ABLIN, D. S. ; GREENSPAN, A. ; REINHART, M. ; GRIX, A. : Differentiation of child abuse from osteogenesis imperfecta. In: *AJR. American journal of roentgenology* 154 (1990), Nr. 5, S. 1035–1046

[4] ACKOFF, R. L.: From data to wisdom. In: *Journal of applied systems analysis* 16 (1989), Nr. 1, S. 3–9

[5] AGRESTI, A. : *Categorical data analysis*. John Wiley & Sons, 2003

[6] ALLIKIVI, M.-L. ; KULL, M. : Non-parametric Bayesian Isotonic Calibration: Fighting Over-Confidence in Binary Classification. In: *Joint European Conference on Machine Learning and Knowledge Discovery in Databases* Springer, 2019, S. 103–120

[7] ALPAYDIN, E. : *Introduction to machine learning*. MIT press, 2020

[8] ALTMAN, D. G.: *Practical statistics for medical research*. CRC press, 1990

[9] AMAT, C. B. ; PERRUCHAS, F. : Evolving cohesion metrics of a research network on rare diseases: a longitudinal study over 14 years. In: *Scientometrics* 108 (2016), Nr. 1, S. 41–56

[10] ANSCOMBE, F. J. ; AUMANN, R. J. u. a.: A definition of subjective probability. In: *Annals of mathematical statistics* 34 (1963), Nr. 1, S. 199–205

[11] AURIA, L. ; MORO, R. A.: Support vector machines (SVM) as a technique for solvency analysis. (2008)

[12] AYMÉ, S. ; SCHMIDTKE, J. : Networking for rare diseases: a necessity for Europe. In: *Bundesgesundheitsblatt-Gesundheitsforschung-Gesundheitsschutz* 50 (2007), Nr. 12, S. 1477–1483

[13] AYODELE, T. O.: Types of machine learning algorithms. In: *New advances in machine learning* 3 (2010), S. 19–48

[14] BALAKRISHNAMA, S. ; GANAPATHIRAJU, A. : Linear discriminant analysis-a brief tutorial. In: *Institute for Signal and information Processing* 18 (1998), S. 1–8

[15] BANERJEE, A. ; JADHAV, S. ; BHAWALKAR, J. : Probability, clinical decision making and hypothesis testing. In: *Industrial psychiatry journal* 18 (2009), Nr. 1, S. 64

[16] BANIKOWSKI, A. K. ; MEHRING, T. A.: Strategies to enhance memory based on brain-research. In: *Focus on Exceptional Children* 32 (1999), Nr. 2, S. 1

[17] BANJANOVIC, E. S. ; OSBORNE, J. W.: Confidence intervals for effect sizes: Applying bootstrap resampling. In: *Practical Assessment, Research, and Evaluation* 21 (2016), Nr. 1, S. 5

[18] BANSAL, V. ; LIBIGER, O. ; TORKAMANI, A. ; SCHORK, N. J.: Statistical analysis strategies for association studies involving rare variants. In: *Nature Reviews Genetics* 11 (2010), Nr. 11, S. 773

[19] BASHEER, I. A. ; HAJMEER, M. : Artificial neural networks: fundamentals, computing, design, and application. In: *Journal of microbiological methods* 43 (2000), Nr. 1, S. 3–31

[20] BAUER, E. ; KOHAVI, R. : An empirical comparison of voting classification algorithms: Bagging, boosting, and variants. In: *Machine learning* 36 (1999), Nr. 1-2, S. 105–139

[21] BELKIN, M. ; HSU, D. ; MA, S. ; MANDAL, S. : Reconciling modern machine-learning practice and the classical bias–variance trade-off. In: *Proceedings of the National Academy of Sciences* 116 (2019), Nr. 32, S. 15849–15854

[22] BELLA, A. ; FERRI, C. ; HERNÁNDEZ-ORALLO, J. ; RAMÍREZ-QUINTANA, M. J.: On the effect of calibration in classifier combination. In: *Applied intelligence* 38 (2013), Nr. 4, S. 566–585

[23] BELLAZZI, R. ; ZUPAN, B. : Predictive data mining in clinical medicine: current issues and guidelines. In: *International journal of medical informatics* 77 (2008), Nr. 2, S. 81–97

[24] BENJAMIN, D. J. ; BERGER, J. O. ; JOHANNESSON, M. ; NOSEK, B. A. ; WAGENMAKERS, E.-J. ; BERK, R. ; BOLLEN, K. A. ; BREMBS, B. ; BROWN, L. ; CAMERER, C. u. a.: Redefine statistical significance. In: *Nature Human Behaviour* 2 (2018), Nr. 1, S. 6–10

[25] BERNARDO, J. ; BAYARRI, M. ; BERGER, J. ; DAWID, A. ; HECKERMAN, D. ; SMITH, A. ; WEST, M. : Generative or discriminative? getting the best of both worlds. In: *Bayesian statistics* 8 (2007), Nr. 3, S. 3–24

[26] BERNER, E. S.: *Clinical decision support systems*. Bd. 233. Springer, 2007

[27] BERTSIMAS, D. ; DUNN, J. : Optimal classification trees. In: *Machine Learning* 106 (2017), Nr. 7, S. 1039–1082

[28] BEST, M. J. ; CHAKRAVARTI, N. : Active set algorithms for isotonic regression; a unifying framework. In: *Mathematical Programming* 47 (1990), Nr. 1-3, S. 425–439

[29] BEYER, K. ; GOLDSTEIN, J. ; RAMAKRISHNAN, R. ; SHAFT, U. : When is "nearest neighbor" meaningful? In: *International conference on database theory* Springer, 1999, S. 217–235

[30] BHATTACHARYYA, S. ; BERKOWITZ, A. L.: Primary angiitis of the central nervous system: avoiding misdiagnosis and missed diagnosis of a rare disease. In: *Practical neurology* 16 (2016), Nr. 3, S. 195–200

[31] BIAU, D. J. ; JOLLES, B. M. ; PORCHER, R. : P value and the theory of hypothesis testing: an explanation for new researchers. In: *Clinical Orthopaedics and Related Research®* 468 (2010), Nr. 3, S. 885–892

[32] BIAU, G. ; SCORNET, E. : A random forest guided tour. In: *Test* 25 (2016), Nr. 2, S. 197–227

[33] BIGGS, N. ; LLOYD, E. K. ; WILSON, R. J.: *Graph Theory, 1736-1936*. Oxford University Press, 1986

[34] BISHOP, C. M.: *Pattern Recognition and Machine Learning*. Springer, 2006

[35] BLÖSS, S. ; KLEMANN, C. ; ROTHER, A.-K. ; MEHMECKE, S. ; SCHUMACHER, U. ; MÜCKE, U. ; MÜCKE, M. ; STIEBER, C. ; KLAWONN, F. ; KORTUM, X. u. a.: Diagnostic needs for rare diseases and shared prediagnostic phenomena: Results of a German-wide expert Delphi survey. In: *PloS one* 12 (2017), Nr. 2, S. e0172532

[36] BÖKEN, B. : On the appropriateness of Platt scaling in classifier calibration. In: *Information Systems* 95 (2021), S. 101641

[37] BOUCKAERT, R. R. ; FRANK, E. : Evaluating the replicability of significance tests for comparing learning algorithms. In: *Pacific-Asia conference on knowledge discovery and data mining* Springer, 2004, S. 3–12

[38] BRACE, I. : *Questionnaire design: How to plan, structure and write survey material for effective market research*. Kogan Page Publishers, 2018

[39] BRADLEY, A. P.: The use of the area under the ROC curve in the evaluation of machine learning algorithms. In: *Pattern recognition* 30 (1997), Nr. 7, S. 1145–1159

[40] BREIMAN, L. : Bagging predictors. In: *Machine learning* 24 (1996), Nr. 2, S. 123–140

[41] BRÖCKER, J. ; SMITH, L. A.: Increasing the reliability of reliability diagrams. In: *Weather and forecasting* 22 (2007), Nr. 3, S. 651–661

[42] BUCKLAND, M. ; GEY, F. : The relationship between recall and precision. In: *Journal of the American society for information science* 45 (1994), Nr. 1, S. 12–19

[43] CAIRO, A. : *The Functional Art: An introduction to information graphics and visualization*. New Riders, 2012

[44] CARUANA, R. ; NICULESCU-MIZIL, A. : An empirical comparison of supervised learning algorithms. In: *Proceedings of the 23rd international conference on Machine learning* ACM, 2006, S. 161–168

[45] CAWLEY, G. C. ; TALBOT, N. L.: Efficient leave-one-out cross-validation of kernel fisher discriminant classifiers. In: *Pattern Recognition* 36 (2003), Nr. 11, S. 2585–2592

[46] CHANDRASHEKAR, G. ; SAHIN, F. : A survey on feature selection methods. In: *Computers & Electrical Engineering* 40 (2014), Nr. 1, S. 16–28

[47] CHAUBEY, Y. P.: *Resampling-based multiple testing: Examples and methods for p-value adjustment*. 1993

[48] CHEN, W. ; SAHINER, B. ; SAMUELSON, F. ; PEZESHK, A. ; PETRICK, N. : Calibration of medical diagnostic classifier scores to the probability of disease. In: *Statistical methods in medical research* 27 (2018), Nr. 5, S. 1394–1409

[49] CHIÒ, A. : ISIS Survey: an international study on the diagnostic process and its implications in amyotrophic lateral sclerosis. In: *Journal of neurology* 246 (1999), Nr. 3, S. III1–III5

[50] CHITTARO, L. : Information visualization and its application to medicine. In: *Artificial intelligence in medicine* 22 (2001), Nr. 2, S. 81–88

[51] CLEVELAND, W. C. ; MCGILL, M. E.: *Dynamic graphics for statistics*. CRC Press, Inc., 1988

[52] CLEVELAND, W. S.: *Visualizing data*. Hobart Press, 1993

[53] COCCHI, M. : *Data fusion methodology and applications*. Elsevier, 2019

[54] COHEN, I. ; GOLDSZMIDT, M. : Properties and benefits of calibrated classifiers. In: *European Conference on Principles of Data Mining and Knowledge Discovery* Springer, 2004, S. 125–136

[55] CONNOLLY, M. P. ; PANDA, S. ; PATRIS, J. ; HAZENBERG, B. P.: Estimating the fiscal impact of rare diseases using a public economic framework: a case study applied to hereditary transthyretin-mediated (hATTR) amyloidosis. In: *Orphanet journal of rare diseases* 14 (2019), Nr. 1, S. 1–9

[56] CONNOLLY, T. M. ; BEGG, C. E.: *Database systems: a practical approach to design, implementation, and management*. Pearson Education, 2005

[57] CORTES, C. ; VAPNIK, V. : Support vector machine. In: *Machine learning* 20 (1995), Nr. 3, S. 273–297

[58] CROOKS, G. E. ; HON, G. ; CHANDONIA, J.-M. ; BRENNER, S. E.: WebLogo: a sequence logo generator. In: *Genome research* 14 (2004), Nr. 6, S. 1188–1190

[59] CRUZ, J. A. ; WISHART, D. S.: Applications of Machine Learning in Cancer Prediction and Prognosis. In: *Cancer Informatics* 2 (2006), S. 117693510600200030. http://dx.doi.org/10.1177/117693510600200030. – DOI 10.1177/117693510600200030

[60] DASH, M. ; LIU, H. : Feature selection for classification. In: *Intelligent data analysis* 1 (1997), Nr. 1-4, S. 131–156

[61] DAVIS, J. ; GOADRICH, M. : The relationship between Precision-Recall and ROC curves. In: *Proceedings of the 23rd international conference on Machine learning*, 2006, S. 233–240

[62] DE LEEUW, J. ; HORNIK, K. ; MAIR, P. : Isotone optimization in R: pool-adjacent-violators algorithm (PAVA) and active set methods. In: *Journal of statistical software* 32 (2010), S. 1–24

[63] DE STEFANO, C. ; DELLA CIOPPA, A. ; MARCELLI, A. : An adaptive weighted majority vote rule for combining multiple classifiers. In: *Object recognition supported by user interaction for service robots* Bd. 2 IEEE, 2002, S. 192–195

[64] DEO, R. C.: Machine learning in medicine. In: *Circulation* 132 (2015), Nr. 20, S. 1920–1930

[65] DEY, K. K. ; XIE, D. ; STEPHENS, M. : A new sequence logo plot to highlight enrichment and depletion. In: *BMC bioinformatics* 19 (2018), Nr. 1, S. 473

[66] DIETTERICH, T. G.: Approximate statistical tests for comparing supervised classification learning algorithms. In: *Neural computation* 10 (1998), Nr. 7, S. 1895–1923

[67] DIETTERICH, T. G.: Ensemble methods in machine learning. In: *International workshop on multiple classifier systems* Springer, 2000, S. 1–15

[68] DIETTERICH, T. G. ; KONG, E. B.: Machine learning bias, statistical bias, and statistical variance of decision tree algorithms / Citeseer. 1995. – Forschungsbericht

[69] DONG, D. ; CHUNG, R. Y.-N. ; CHAN, R. H. ; GONG, S. ; XU, R. H.: Why is misdiagnosis more likely among some people with rare diseases than others? Insights from a population-based cross-sectional study in China. In: *Orphanet journal of rare diseases* 15 (2020), Nr. 1, S. 1–12

[70] DUA, D. ; GRAFF, C. : *UCI Machine Learning Repository*. http://archive.ics.uci.edu/ml. Version: 2017

[71] DUDANI, S. A.: The distance-weighted k-nearest-neighbor rule. In: *IEEE Transactions on Systems, Man, and Cybernetics* (1976), Nr. 4, S. 325–327

[72] DŽEROSKI, S. ; ŽENKO, B. : Is combining classifiers with stacking better than selecting the best one? In: *Machine learning* 54 (2004), Nr. 3, S. 255–273

[73] EFRON, B. ; TIBSHIRANI, R. J.: *An introduction to the bootstrap*. CRC press, 1994

[74] EURORDIS, A. K. ; FAURISSON, F. : *The Voice of 12,000 Patients. Experiences and Expectations of Rare Disease Patients on Diagnosis and Care in Europe.* EURORDIS-Rare Diseases Eu, 2009

[75] EVERITT, B. S.: *Statistical methods for medical investigations.* Edward Arnold London, 1994

[76] EZEKIEL, M. : Methods of correlation analysis. (1930)

[77] FAWCETT, T. : ROC graphs: Notes and practical considerations for researchers. In: *Machine learning* 31 (2004), Nr. 1, S. 1–38

[78] FAWCETT, T. ; NICULESCU-MIZIL, A. : PAV and the ROC convex hull. In: *Machine Learning* 68 (2007), Nr. 1, S. 97–106

[79] FAYYAD, U. ; PIATETSKY-SHAPIRO, G. ; SMYTH, P. : From data mining to knowledge discovery in databases. In: *AI magazine* 17 (1996), Nr. 3, S. 37

[80] FAYYAD, U. M. ; PIATETSKY-SHAPIRO, G. ; SMYTH, P. u. a.: Knowledge Discovery and Data Mining: Towards a Unifying Framework. In: *KDD* Bd. 96, 1996, S. 82–88

[81] FERNÁNDEZ-DELGADO, M. ; CERNADAS, E. ; BARRO, S. ; AMORIM, D. : Do we need hundreds of classifiers to solve real world classification problems? In: *The Journal of Machine Learning Research* 15 (2014), Nr. 1, S. 3133–3181

[82] FOSTER, K. R. ; KOPROWSKI, R. ; SKUFCA, J. D.: Machine learning, medical diagnosis, and biomedical engineering research-commentary. In: *Biomedical engineering online* 13 (2014), Nr. 1, S. 94

[83] FREUND, Y. ; SCHAPIRE, R. E. u. a.: Experiments with a new boosting algorithm. In: *Icml* Bd. 96 Citeseer, 1996, S. 148–156

[84] FRIENDLY, M. : A brief history of data visualization. In: *Handbook of data visualization.* Springer, 2008, S. 15–56

[85] GARCÍA-PEDRAJAS, N. ; ORTIZ-BOYER, D. : An empirical study of binary classifier fusion methods for multiclass classification. In: *Information Fusion* 12 (2011), Nr. 2, S. 111–130

[86] GASHLER, M. ; GIRAUD-CARRIER, C. ; MARTINEZ, T. : Decision tree ensemble: Small heterogeneous is better than large homogeneous. In: *2008 Seventh International Conference on Machine Learning and Applications* IEEE, 2008, S. 900–905

[87] GEHLENBORG, N. ; WONG, B. : *Heat maps.* 2012

[88] GEMAN, S. ; BIENENSTOCK, E. ; DOURSAT, R. : Neural networks and the bias/variance dilemma. In: *Neural computation* 4 (1992), Nr. 1, S. 1–58

[89] GILPIN, L. H. ; BAU, D. ; YUAN, B. Z. ; BAJWA, A. ; SPECTER, M. ; KAGAL, L. : Explaining explanations: An overview of interpretability of machine learning. In: *2018 IEEE 5th International Conference on data science and advanced analytics (DSAA)* IEEE, 2018, S. 80–89

[90] GOODHUE, D. L. ; WYBO, M. D. ; KIRSCH, L. J.: The impact of data integration on the costs and benefits of information systems. In: *MiS Quarterly* (1992), S. 293–311

[91] GREENES, R. A.: *Clinical decision support: the road ahead*. Elsevier, 2011

[92] GRIGULL, L. ; LECHNER, W. ; PETRI, S. ; KOLLEWE, K. ; DENGLER, R. ; MEHMECKE, S. ; SCHUMACHER, U. ; LÜCKE, T. ; SCHNEIDER-GOLD, C. ; KÖHLER, C. u. a.: Diagnostic support for selected neuromuscular diseases using answer-pattern recognition and data mining techniques: a proof of concept multicenter prospective trial. In: *BMC medical informatics and decision making* 16 (2016), Nr. 1, S. 31

[93] GRIGULL, L. ; MEHMECKE, S. ; ROTHER, A.-K. ; BLÖSS, S. ; KLEMANN, C. ; SCHUMACHER, U. ; MÜCKE, U. ; KORTUM, X. ; LECHNER, W. ; KLAWONN, F. u. a.: Common pre-diagnostic features in individuals with different rare diseases represent a key for diagnostic support with computerized pattern recognition? In: *PloS one* 14 (2019), Nr. 10, S. 1–15

[94] GRINSTEIN, U. M. F. G. G. ; WIERSE, A. : *Information visualization in data mining and knowledge discovery*. Morgan Kaufmann, 2002

[95] GUPTA, S. ; GUPTA, A. : Dealing with noise problem in machine learning data-sets: A systematic review. In: *Procedia Computer Science* 161 (2019), S. 466–474

[96] GUYON, I. ; ELISSEEFF, A. : An introduction to variable and feature selection. In: *Journal of machine learning research* 3 (2003), Nr. Mar, S. 1157–1182

[97] HAMARASHID, H. K.: Utilizing statistical tests for comparing machine learning algorithms. In: *Kurd J Appl Res* 6 (2021), Nr. 1, S. 69–74

[98] HAN, J. ; PEI, J. ; KAMBER, M. : *Data mining: concepts and techniques*. Elsevier, 2011

[99] HAND, D. J.: Principles of data mining. In: *Drug safety* 30 (2007), Nr. 7, S. 621–622

[100] HANLEY, J. A. ; MCNEIL, B. J.: The meaning and use of the area under a receiver operating characteristic (ROC) curve. In: *Radiology* 143 (1982), Nr. 1, S. 29–36

[101] HARDOON, D. R. ; SZEDMAK, S. ; SHAWE-TAYLOR, J. : Canonical correlation analysis: An overview with application to learning methods. In: *Neural computation* 16 (2004), Nr. 12, S. 2639–2664

[102] HARSHVARDHAN, G. ; GOURISARIA, M. K. ; PANDEY, M. ; RAUTARAY, S. S.: A comprehensive survey and analysis of generative models in machine learning. In: *Computer Science Review* 38 (2020), S. 100285

[103] HASSON, F. ; KEENEY, S. ; MCKENNA, H. : Research guidelines for the Delphi survey technique. In: *Journal of advanced nursing* 32 (2000), Nr. 4, S. 1008–1015

[104] HASSOUN, M. H. u. a.: *Fundamentals of artificial neural networks*. MIT press, 1995

[105] HASTIE, T. ; TIBSHIRANI, R. ; FRIEDMAN, J. : Unsupervised learning. In: *The elements of statistical learning*. Springer, 2009, S. 485–585

[106] HAYNES, R. B. ; SACKETT, D. L. ; RICHARDSON, W. S. ; ROSENBERG, W. ; LANGLEY, G. R.: Evidence-based medicine: How to practice & teach EBM. In: *Canadian Medical Association. Journal* 157 (1997), Nr. 6, S. 788

[107] HEIBERGER, R. M. ; ROBBINS, N. B.: Design of diverging stacked bar charts for likert scales and other applications. In: *Journal of Statistical Software* 57 (2014), Nr. 5, S. 1–32

[108] HENZI, A. ; MÖSCHING, A. ; DÜMBGEN, L. : Accelerating the pool-adjacent-violators algorithm for isotonic distributional regression. In: *Methodology and computing in applied probability* (2022), S. 1–13

[109] HERNÁNDEZ, M. A. ; STOLFO, S. J.: Real-world data is dirty: Data cleansing and the merge/purge problem. In: *Data mining and knowledge discovery* 2 (1998), Nr. 1, S. 9–37

[110] HO, T. K. ; HULL, J. J. ; SRIHARI, S. N.: Decision combination in multiple classifier systems. In: *Pattern Analysis and Machine Intelligence, IEEE Transactions on* 16 (1994), Nr. 1, S. 66–75

[111] HOFFMAN, M. M. ; BUSKE, O. J. ; WANG, J. ; WENG, Z. ; BILMES, J. A. ; NOBLE, W. S.: Unsupervised pattern discovery in human chromatin structure through genomic segmentation. In: *Nature methods* 9 (2012), Nr. 5, S. 473

[112] HOSSIN, M. ; SULAIMAN, M. : A review on evaluation metrics for data classification evaluations. In: *International Journal of Data Mining & Knowledge Management Process* 5 (2015), Nr. 2, S. 1

[113] HUYN, N. ; MELMON, K. ; PERRONE, A. : *Computerized clinical questionnaire with dynamically presented questions*. Jul. 20 2001. – US Patent App. 09/910,463

[114] IJ, H. : Statistics versus machine learning. In: *Nat Methods* 15 (2018), Nr. 4, S. 233

[115] ILYASOVA, N. Y. ; KUPRIYANOV, A. ; PARINGER, R. : The discriminant analysis application to refine the diagnostic features of blood vessels images. In: *Optical Memory and Neural Networks* 24 (2015), Nr. 4, S. 309–313

[116] IZENMAN, A. J.: Modern multivariate statistical techniques. In: *Regression, classification and manifold learning* 10 (2008), S. 978–0

[117] JABBAR, H. ; KHAN, R. Z.: Methods to avoid over-fitting and under-fitting in supervised machine learning (comparative study). In: *Computer Science, Communication and Instrumentation Devices* 70 (2015)

[118] JAFFAR, M. A.: Deep Learning based Computer Aided Diagnosis System for Breast Mammograms. In: *International Journal of Advanced Computer Science and Applications* 8 (2017), Nr. 7

[119] JAIN, A. K. ; DUIN, R. P. ; MAO, J. : Statistical pattern recognition: A review. In: *Pattern Analysis and Machine Intelligence, IEEE Transactions on* 22 (2000), Nr. 1, S. 4–37

[120] JAIN, A. K. ; MAO, J. ; MOHIUDDIN, K. M.: Artificial neural networks: A tutorial. In: *Computer* 29 (1996), Nr. 3, S. 31–44

[121] JAMES, G. ; WITTEN, D. ; HASTIE, T. ; TIBSHIRANI, R. : *An introduction to statistical learning*. Bd. 112. Springer, 2013

[122] JANSEN, A. C. ; AALST-COHEN, E. S. ; HUTTEN, B. A. ; BÜLLER, H. R. ; KASTELEIN, J. J. ; PRINS, M. H.: Guidelines were developed for data collection from medical records for use in retrospective analyses. In: *Journal of clinical epidemiology* 58 (2005), Nr. 3, S. 269–274

[123] JIANG, X. ; OSL, M. ; KIM, J. ; OHNO-MACHADO, L. : Smooth isotonic regression: A new method to calibrate predictive models. In: *AMIA Summits on Translational Science Proceedings* 2011 (2011), S. 16

[124] JIANG, Z. ; LIU, H. : Ensemble learning based on probability calibration. In: *Journal of Computer Applications* 36 (2016), Nr. 2, S. 291. `http://dx.doi.org/10.11772/j.issn.1001-9081.2016.02.0291`. – DOI 10.11772/j.issn.1001–9081.2016.02.0291

[125] JOHN, G. H. ; LANGLEY, P. : Estimating continuous distributions in Bayesian classifiers. In: *arXiv preprint arXiv:1302.4964* (2013)

[126] JOLLIFFE, I. : *Principal component analysis*. Springer, 2011

[127] JUNG, Y. : Multiple predicting K-fold cross-validation for model selection. In: *Journal of Nonparametric Statistics* 30 (2018), Nr. 1, S. 197–215

[128] KALAI, A. T. ; SASTRY, R. : The Isotron Algorithm: High-Dimensional Isotonic Regression. In: *COLT* Citeseer, 2009

[129] KALAPANIDAS, E. ; AVOURIS, N. ; CRACIUN, M. ; NEAGU, D. : Machine learning algorithms: a study on noise sensitivity. In: *Proc. 1st Balcan Conference in Informatics*, 2003, S. 356–365

[130] KANJI, G. K.: *100 statistical tests*. Sage, 2006

[131] KELLER, J. M. ; GRAY, M. R. ; GIVENS, J. A.: A fuzzy k-nearest neighbor algorithm. In: *IEEE transactions on systems, man, and cybernetics* (1985), Nr. 4, S. 580–585

[132] KIERNAN, M. C. ; VUCIC, S. ; CHEAH, B. C. ; TURNER, M. R. ; EISEN, A. ; HARDIMAN, O. ; BURRELL, J. R. ; ZOING, M. C.: Amyotrophic lateral sclerosis. In: *The lancet* 377 (2011), Nr. 9769, S. 942–955

[133] KINNEY, J. B. ; ATWAL, G. S.: Equitability, mutual information, and the maximal information coefficient. In: *Proceedings of the National Academy of Sciences* 111 (2014), Nr. 9, S. 3354–3359

[134] KLAWONN, F. ; LECHNER, W. ; GRIGULL, L. : Case-centred multidimensional scaling for classification visualisation in medical diagnosis. In: *International Conference on Health Information Science* Springer, 2013, S. 137–148

[135] KOHAVI, R. : Scaling up the accuracy of Naive-Bayes classifiers: a decision-tree hybrid. In: *KDD* Bd. 96 Citeseer, 1996, S. 202–207

[136] KOHAVI, R. u. a.: A study of cross-validation and bootstrap for accuracy estimation and model selection. In: *Ijcai* Bd. 14 Montreal, Canada, 1995, S. 1137–1145

[137] KOLODNER, J. L.: An introduction to case-based reasoning. In: *Artificial intelligence review* 6 (1992), Nr. 1, S. 3–34

[138] KONONENKO, I. : Machine learning for medical diagnosis: history, state of the art and perspective. In: *Artificial Intelligence in medicine* 23 (2001), Nr. 1, S. 89–109

[139] KORTUM, X. ; GRIGULL, L. ; LECHNER, W. ; KLAWONN, F. : A Dynamic Adaptive Questionnaire for Improved Disease Diagnostics. In: *International Symposium on Intelligent Data Analysis* Springer, 2017, S. 162–172

[140] KORTUM, X. ; GRIGULL, L. ; MUECKE, U. ; LECHNER, W. ; KLAWONN, F. : Diagnosis Support for Orphan Diseases: A Case Study Using a Classifier Fusion Method. In: *International Conference on Intelligent Data Engineering and Automated Learning* Springer International Publishing, 2016, S. 379–385

[141] KORTUM, X. ; GRIGULL, L. ; MUECKE, U. ; LECHNER, W. ; KLAWONN, F. : Improving the Decision Support in Diagnostic Systems using Classifier Probability Calibration. In: *International Conference on Intelligent Data Engineering and Automated Learning* Springer, 2018, S. 162–172

[142] KOSSLYN, S. M.: *Image and brain: The resolution of the imagery debate*. MIT press, 1996

[143] KOZAK, A. ; KOZAK, R. : Does cross validation provide additional information in the evaluation of regression models? In: *Canadian Journal of Forest Research* 33 (2003), Nr. 6, S. 976–987

[144] KRSTAJIC, D. ; BUTUROVIC, L. J. ; LEAHY, D. E. ; THOMAS, S. : Cross-validation pitfalls when selecting and assessing regression and classification models. In: *Journal of cheminformatics* 6 (2014), Nr. 1, S. 1–15

[145] KUHN, M. ; JOHNSON, K. u. a.: *Applied predictive modeling*. Bd. 26. Springer, 2013

[146] KULLBACK, S. : *Information theory and statistics*. Courier Corporation, 1997

[147] KULLBACK, S. ; LEIBLER, R. A.: On information and sufficiency. In: *The annals of mathematical statistics* 22 (1951), Nr. 1, S. 79–86

[148] KUMAR, R. ; INDRAYAN, A. : Receiver operating characteristic (ROC) curve for medical researchers. In: *Indian pediatrics* 48 (2011), Nr. 4, S. 277–287

[149] KUNCHEVA, L. I.: A theoretical study on six classifier fusion strategies. In: *IEEE Transactions on Pattern Analysis & Machine Intelligence* (2002), Nr. 2, S. 281–286

[150] KUNCHEVA, L. I.: *Combining pattern classifiers: methods and algorithms*. John Wiley & Sons, 2014

[151] KUNCHEVA, L. I. ; BEZDEK, J. C. ; DUIN, R. P.: Decision templates for multiple classifier fusion: an experimental comparison. In: *Pattern recognition* 34 (2001), Nr. 2, S. 299–314

[152] KUNCHEVA, L. I. ; WHITAKER, C. J.: Measures of diversity in classifier ensembles and their relationship with the ensemble accuracy. In: *Machine learning* 51 (2003), Nr. 2, S. 181–207

[153] KUO, W.-J. ; CHANG, R.-F. ; CHEN, D.-R. ; LEE, C. C.: Data mining with decision trees for diagnosis of breast tumor in medical ultrasonic images. In: *Breast cancer research and treatment* 66 (2001), Nr. 1, S. 51–57

[154] KUO, W.-J. ; CHANG, R.-F. ; MOON, W. K. ; LEE, C. C. ; CHEN, D.-R. : Computer-aided diagnosis of breast tumors with different US systems. In: *Academic radiology* 9 (2002), Nr. 7, S. 793–799

[155] KWON, B. C. ; CHOI, M.-J. ; KIM, J. T. ; CHOI, E. ; KIM, Y. B. ; KWON, S. ; SUN, J. ; CHOO, J. : Retainvis: Visual analytics with interpretable and interactive recurrent neural networks on electronic medical records. In: *IEEE transactions on visualization and computer graphics* 25 (2018), Nr. 1, S. 299–309

[156] LANDAIS, P. ; MESSIAEN, C. ; RATH, A. ; LE MIGNOT, L. ; DUFOUR, E. ; SAID, M. B. ; JAÏS, J. P. ; TOUBIANA, L. ; BAUJAT, G. ; BOURDON-LANOY, E. u. a.: CEMARA an information system for rare diseases. In: *MedInfo*, 2010, S. 481–485

[157] LAWYER, T. ; NETSKY, M. G.: Amyotrophic lateral sclerosis: a clinicoanatomic study of fifty-three cases. In: *AMA Archives of Neurology & Psychiatry* 69 (1953), Nr. 2, S. 171–192

[158] LE CAM, Y. ; TEJADA, P. : Rare diseases A hidden priority. In: *Organisation for Economic Cooperation and Development. The OECD Observer* (2010), Nr. 281, S. 18

[159] LEONARD, D. ; SENSIPER, S. : The role of tacit knowledge in group innovation. In: *California management review* 40 (1998), Nr. 3, S. 112–132

[160] LIAW, A. ; WIENER, M. : Classification and regression by Random Forest. In: *R news* 2 (2002), Nr. 3, S. 18–22

[161] LIN, J. : Divergence measures based on the Shannon entropy. In: *IEEE Transactions on Information theory* 37 (1991), Nr. 1, S. 145–151

[162] LIU, H. ; MOTODA, H. : *Computational methods of feature selection*. CRC Press, 2007

[163] LIU, Z. ; YAO, C. ; YU, H. ; WU, T. : Deep reinforcement learning with its application for lung cancer detection in medical Internet of Things. In: *Future Generation Computer Systems* 97 (2019), S. 1–9

[164] LOWRY, R. : Concepts and applications of inferential statistics. (2014)

[165] LU, Y. : Knowledge integration in a multiple classifier system. In: *Applied Intelligence* 6 (1996), Nr. 2, S. 75–86

[166] MA, L. ; LIU, X. ; SONG, L. ; ZHOU, C. ; ZHAO, X. ; ZHAO, Y. : A new classifier fusion method based on historical and on-line classification reliability for recognizing common CT imaging signs of lung diseases. In: *Computerized Medical Imaging and Graphics* 40 (2015), S. 39–48

[167] MADABHUSHI, A. ; AGNER, S. ; BASAVANHALLY, A. ; DOYLE, S. ; LEE, G. : Computer-aided prognosis: predicting patient and disease outcome via quantitative fusion of multi-scale, multi-modal data. In: *Computerized medical imaging and graphics* 35 (2011), Nr. 7

[168] MAHMUD, M. ; KAISER, M. S. ; HUSSAIN, A. ; VASSANELLI, S. : Applications of deep learning and reinforcement learning to biological data. In: *IEEE transactions on neural networks and learning systems* 29 (2018), Nr. 6, S. 2063–2079

[169] MAIR, P. ; HORNIK, K. ; LEEUW, J. de: Isotone optimization in R: pool-adjacent-violators algorithm (PAVA) and active set methods. In: *Journal of statistical software* 32 (2009), Nr. 5, S. 1–24

[170] MANGASARIAN, O. L. ; WOLBERG, W. H.: Cancer diagnosis via linear programming / University of Wisconsin-Madison Department of Computer Sciences. 1990. – Forschungsbericht

[171] MANSOURYPOOR, F. ; ASADI, S. : Development of a reinforcement learning-based evolutionary fuzzy rule-based system for diabetes diagnosis. In: *Computers in Biology and Medicine* 91 (2017), S. 337–352

[172] MARLIN, B. M. ; KALE, D. C. ; KHEMANI, R. G. ; WETZEL, R. C.: Unsupervised pattern discovery in electronic health care data using probabilistic clustering models. In: *Proceedings of the 2nd ACM SIGHIT international health informatics symposium*, 2012, S. 389–398

[173] MARSLAND, S. : *Machine learning: an algorithmic perspective*. Chapman and Hall/CRC, 2011

[174] MCAULIFFE, M. J. ; LALONDE, F. M. ; MCGARRY, D. ; GANDLER, W. ; CSAKY, K. ; TRUS, B. L.: Medical image processing, analysis and visualization in clinical research. In: *Computer-Based Medical Systems, 2001. CBMS 2001. Proceedings. 14th IEEE Symposium on* IEEE, 2001, S. 381–386

[175] MCHUGH, M. L.: The chi-square test of independence. In: *Biochemia medica* 23 (2013), Nr. 2, S. 143–149

[176] MEEK, C. : A characterization of prediction errors. In: *arXiv preprint arXiv:1611.05955* (2016)

[177] MEIER, P. : Variance of a weighted mean. In: *Biometrics* 9 (1953), Nr. 1, S. 59–73

[178] MELNIKOVA, I. : *Rare diseases and orphan drugs*. 2012

[179] MICHALSKI, R. S. ; CARBONELL, J. G. ; MITCHELL, T. M.: *Machine learning: An artificial intelligence approach*. Springer Science & Business Media, 2013

[180] MIKA, S. ; RATSCH, G. ; WESTON, J. ; SCHOLKOPF, B. ; MULLERS, K.-R. : Fisher discriminant analysis with kernels. In: *Neural networks for signal processing IX, 1999. Proceedings of the 1999 IEEE signal processing society workshop.* Ieee, 1999, S. 41–48

[181] MITCHELL, T. M.: Machine learning and data mining. In: *Communications of the ACM* 42 (1999), Nr. 11, S. 30–36

[182] MOHAMED, W. N. H. W. ; SALLEH, M. N. M. ; OMAR, A. H.: A comparative study of reduced error pruning method in decision tree algorithms. In: *2012 IEEE International conference on control system, computing and engineering* IEEE, 2012, S. 392–397

[183] MORENO-SECO, F. ; INESTA, J. M. ; DE LEÓN, P. J. P. ; MICÓ, L. : Comparison of classifier fusion methods for classification in pattern recognition tasks. In: *Joint IAPR International Workshops on Statistical Techniques in Pattern Recognition (SPR) and Structural and Syntactic Pattern Recognition (SSPR)* Springer, 2006, S. 705–713

[184] MORTON, A. B. ; MAREELS, I. M.: An efficient brute-force solution to the network reconfiguration problem. In: *IEEE Transactions on Power Delivery* 15 (2000), Nr. 3, S. 996–1000

[185] MÜCKE, U. ; KLEMANN, C. ; BAUMANN, U. ; MEYER-BAHLBURG, A. ; KORTUM, X. ; KLAWONN, F. ; LECHNER, W. M. ; GRIGULL, L. : Patient's experience in Pediatric Primary Immunodeficiency Disorders: computerized classification of Questionnaires. In: *Frontiers in immunology* 8 (2017), S. 384

[186] MUSEN, M. A. ; MIDDLETON, B. ; GREENES, R. A.: Clinical decision-support systems. In: *Biomedical informatics*. Springer, 2014, S. 643–674

[187] NASRABADI, N. M.: Pattern recognition and machine learning. In: *Journal of electronic imaging* 16 (2007), Nr. 4, S. 049901

[188] NEMOTO, T. ; BEGLAR, D. : Likert-scale questionnaires. In: *JALT 2013 conference proceedings*, 2014, S. 1–8

[189] NESSE, R. M. ; WILLIAMS, G. C.: *Why we get sick: The new science of Darwinian medicine*. Vintage, 2012

[190] NG, A. ; JORDAN, M. : On discriminative vs. generative classifiers: A comparison of logistic regression and naive bayes. In: *Advances in neural information processing systems* 14 (2001)

[191] NICULESCU-MIZIL, A. ; CARUANA, R. : Predicting good probabilities with supervised learning. In: *Proceedings of the 22nd international conference on Machine learning* ACM, 2005, S. 625–632

[192] OBERMEYER, Z. ; EMANUEL, E. J.: Predicting the future—big data, machine learning, and clinical medicine. In: *The New England journal of medicine* 375 (2016), Nr. 13, S. 1216

[193] OLSON, D. L. ; DELEN, D. : *Advanced data mining techniques*. Springer Science & Business Media, 2008

[194] OPITZ, D. ; MACLIN, R. : Popular ensemble methods: An empirical study. In: *Journal of artificial intelligence research* 11 (1999), S. 169–198

[195] ORGANIZATION, W. H.: *Social, public health, human rights, ethical and legal implications of misdiagnosis of HIV status*. `http://www.who.int/hiv/pub/meetingreports/hiv-misdiagnosis-report/en/`. Version: 2016

[196] ORGANIZATION, W. H.: *Cancer in Children*. `http://www.who.int/news-room/fact-sheets/detail/cancer-in-children`. Version: 2018

[197] OSHIRO, T. M. ; PEREZ, P. S. ; BARANAUSKAS, J. A.: How many trees in a random forest? In: *International workshop on machine learning and data mining in pattern recognition* Springer, 2012, S. 154–168

[198] OSISANWO, F. ; AKINSOLA, J. ; AWODELE, O. ; HINMIKAIYE, J. ; OLAKANMI, O. ; AKINJOBI, J. : Supervised machine learning algorithms: classification and comparison. In: *International Journal of Computer Trends and Technology (IJCTT)* 48 (2017), Nr. 3, S. 128–138

[199] OZCIFT, A. ; GULTEN, A. : Classifier ensemble construction with rotation forest to improve medical diagnosis performance of machine learning algorithms. In: *Computer methods and programs in biomedicine* 104 (2011), Nr. 3, S. 443–451

[200] PALANIAPPAN, S. ; AWANG, R. : Intelligent heart disease prediction system using data mining techniques. In: *Computer Systems and Applications, 2008. AICCSA 2008. IEEE/ACS International Conference on* IEEE, 2008, S. 108–115

[201] PANINSKI, L. : Estimation of entropy and mutual information. In: *Neural computation* 15 (2003), Nr. 6, S. 1191–1253

[202] PAVAN, S. ; ROMMEL, K. ; MATEO MARQUINA, M. E. ; HÖHN, S. ; LANNEAU, V. ; RATH, A. : Clinical practice guidelines for rare diseases: the orphanet database. In: *PloS one* 12 (2017), Nr. 1, S. e0170365

[203] PEDREGOSA, F. ; VAROQUAUX, G. ; GRAMFORT, A. ; MICHEL, V. ; THIRION, B. ; GRISEL, O. ; BLONDEL, M. ; PRETTENHOFER, P. ; WEISS, R. ; DUBOURG, V. ; VANDERPLAS, J. ; PASSOS, A. ; COURNAPEAU, D. ; BRUCHER, M. ; PERROT, M. ; DUCHESNAY, E. : Scikit-learn: Machine Learning in Python. In: *Journal of Machine Learning Research* 12 (2011), S. 2825–2830

[204] PEROTTE, A. ; PIVOVAROV, R. ; NATARAJAN, K. ; WEISKOPF, N. ; WOOD, F. ; ELHADAD, N. : Diagnosis code assignment: models and evaluation metrics. In: *Journal of the American Medical Informatics Association* 21 (2014), Nr. 2, S. 231–237

[205] PETRIE, A. ; SABIN, C. : *Medical statistics at a glance*. John Wiley & Sons, 2019

[206] PLATT, J. u. a.: Probabilistic outputs for support vector machines and comparisons to regularized likelihood methods. In: *Advances in large margin classifiers* 10 (1999), Nr. 3, S. 61–74

[207] POHAR, M. ; BLAS, M. ; TURK, S. : Comparison of logistic regression and linear discriminant analysis: a simulation study. In: *Metodoloski zvezki* 1 (2004), Nr. 1, S. 143

[208] POLAT, K. ; GÜNEŞ, S. ; ARSLAN, A. : A cascade learning system for classification of diabetes disease: Generalized discriminant analysis and least square support vector machine. In: *Expert systems with applications* 34 (2008), Nr. 1, S. 482–487

[209] POLKINGHORNE, D. E.: Phenomenological research methods. In: *Existential-phenomenological perspectives in psychology*. Springer, 1989, S. 41–60

[210] PRATHER, J. C. ; LOBACH, D. F. ; GOODWIN, L. K. ; HALES, J. W. ; HAGE, M. L. ; HAMMOND, W. E.: Medical data mining: knowledge discovery in a clinical data warehouse. In: *Proceedings of the AMIA annual fall symposium* American Medical Informatics Association, 1997, S. 101

[211] PREIM, B. ; BARTZ, D. : *Visualization in medicine: theory, algorithms, and applications*. Elsevier, 2007

[212] QI, Y. : Random forest for bioinformatics. In: *Ensemble machine learning*. Springer, 2012, S. 307–323

[213] RAGHAVAN, V. ; BOLLMANN, P. ; JUNG, G. S.: A critical investigation of recall and precision as measures of retrieval system performance. In: *ACM Transactions on Information Systems (TOIS)* 7 (1989), Nr. 3, S. 205–229

[214] RAJKOMAR, A. ; DEAN, J. ; KOHANE, I. : Machine learning in medicine. In: *New England Journal of Medicine* 380 (2019), Nr. 14, S. 1347–1358

[215] RASCHKA, S. : Model evaluation, model selection, and algorithm selection in machine learning. In: *arXiv preprint arXiv:1811.12808* (2018)

[216] RATH, A. ; OLRY, A. ; DHOMBRES, F. ; BRANDT, M. M. ; URBERO, B. ; AYME, S. : Representation of rare diseases in health information systems: the Orphanet approach to serve a wide range of end users. In: *Human mutation* 33 (2012), Nr. 5, S. 803–808

[217] RAYMOND, M. ; ROUSSET, F. : An exact test for population differentiation. In: *Evolution* 49 (1995), Nr. 6, S. 1280–1283

[218] REGULATION, O. M. P.: Regulation (EC) No 141/2000 of the European Parliament and of the Council of 16 December 1999 on orphan medicinal products. In: *Off J* 18 (2000), S. 15

[219] RESHEF, D. N. ; RESHEF, Y. A. ; FINUCANE, H. K. ; GROSSMAN, S. R. ; MCVEAN, G. ; TURNBAUGH, P. J. ; LANDER, E. S. ; MITZENMACHER, M. ; SABETI, P. C.: Detecting novel associations in large data sets. In: *science* 334 (2011), Nr. 6062, S. 1518–1524

[220] RESHEF, D. N. ; RESHEF, Y. A. ; SABETI, P. C. ; MITZENMACHER, M. M.: An empirical study of leading measures of dependence. In: *arXiv preprint arXiv:1505.02214* (2015)

[221] REZAEI, K. ; AGAHI, H. ; MAHMOODZADEH, A. : A Weighted Voting Classifiers Ensemble for the Brain Tumors Classification in MR Images. In: *IETE Journal of Research* (2020), S. 1–14

[222] RISH, I. u. a.: An empirical study of the naive Bayes classifier. In: *IJCAI 2001 workshop on empirical methods in artificial intelligence* Bd. 3, 2001, S. 41–46

[223] RODRIGUES, L. ; KIRKWOOD, B. R.: Case-control designs in the study of common diseases: updates on the demise of the rare disease assumption and the choice of sampling scheme for controls. In: *International journal of epidemiology* 19 (1990), Nr. 1, S. 205–213

[224] ROKACH, L. : Ensemble methods for classifiers. In: *Data mining and knowledge discovery handbook*. Springer, 2005, S. 957–980

[225] RONICKE, S. ; HIRSCH, M. C. ; TÜRK, E. ; LARIONOV, K. ; TIENTCHEU, D. ; WAGNER, A. D.: Can a decision support system accelerate rare disease diagnosis? Evaluating the potential impact of Ada DX in a retrospective study. In: *Orphanet journal of rare diseases* 14 (2019), Nr. 1, S. 1–12

[226] RUSTEMPASIC, I. ; CAN, M. : Diagnosis of parkinson's disease using fuzzy c-means clustering and pattern recognition. In: *Southeast Europe Journal of Soft Computing* 2 (2013), Nr. 1

[227] RUTA, D. ; GABRYS, B. : An overview of classifier fusion methods. In: *Computing and Information systems* 7 (2000), Nr. 1, S. 1–10

[228] SACKETT, D. L. ; ROSENBERG, W. M. ; GRAY, J. M. ; HAYNES, R. B. ; RICHARDSON, W. S.: *Evidence based medicine: what it is and what it isn't*. 1996

[229] SAITO, T. ; REHMSMEIER, M. : The precision-recall plot is more informative than the ROC plot when evaluating binary classifiers on imbalanced datasets. In: *PloS one* 10 (2015), Nr. 3, S. e0118432

[230] SCHAFFER, C. : A conservation law for generalization performance. In: *Machine Learning Proceedings 1994*. Elsevier, 1994, S. 259–265

[231] SCHIEPPATI, A. ; HENTER, J.-I. ; DAINA, E. ; APERIA, A. : Why rare diseases are an important medical and social issue. In: *The Lancet* 371 (2008), Nr. 9629, S. 2039–2041

[232] SCHNEIDER, T. D. ; STEPHENS, R. M.: Sequence logos: a new way to display consensus sequences. In: *Nucleic acids research* 18 (1990), Nr. 20, S. 6097–6100

[233] SCHNOTZ, W. : An integrated model of text and picture comprehension. In: *The Cambridge handbook of multimedia learning* 49 (2005), S. 69

[234] SHAFFER, J. P.: Multiple hypothesis testing. In: *Annual review of psychology* 46 (1995), Nr. 1, S. 561–584

[235] SHANNON, C. E.: A mathematical theory of communication. In: *ACM SIGMOBILE mobile computing and communications review* 5 (2001), Nr. 1, S. 3–55

[236] SHAO, F. ; LIU, H. : The Theoretical and Experimental Analysis of the Maximal Information Coefficient Approximate Algorithm. In: *Journal of Systems Science and Information* 9 (2021), Nr. 1, S. 95–104

[237] SHOSHA, G. A.: Employment of Colaizzi's strategy in descriptive phenomenology: A reflection of a researcher. In: *European Scientific Journal* 8 (2012), Nr. 27

[238] SIEG, A.-L. ; DAS, A. M. ; MUSCHOL, N. M. ; KÖHN, A. ; LAMPE, C. ; KORTUM, X. ; MEHMECKE, S. ; BLÖSS, S. ; LECHNER, W. ; KLAWONN, F. u. a.: Künstliche Intelligenz zur diagnostischen Unterstützung ausgewählter seltener lysosomaler Speichererkrankungen: Ergebnisse einer Pilotstudie. In: *Klinische Pädiatrie* 231 (2019), Nr. 02, S. 60–66

[239] SIMON, S. D.: *Statistical evidence in medical trials: what do the data really tell us?* Oxford University Press, 2006

[240] SINCERO, S. M.: *Types of Survey Questions*. `https://explorable.com/types-of-survey-questions`. Version: April 2012

[241] SINGH, A. ; THAKUR, N. ; SHARMA, A. : A review of supervised machine learning algorithms. In: *2016 3rd International Conference on Computing for Sustainable Global Development (INDIACom)* Ieee, 2016, S. 1310–1315

[242] SOKOLOVA, M. ; LAPALME, G. : A systematic analysis of performance measures for classification tasks. In: *Information processing & management* 45 (2009), Nr. 4, S. 427–437

[243] SONI, J. ; ANSARI, U. ; SHARMA, D. ; SONI, S. : Predictive data mining for medical diagnosis: An overview of heart disease prediction. In: *International Journal of Computer Applications* 17 (2011), Nr. 8, S. 43–48

[244] STEVENS, S. S. u. a.: On the theory of scales of measurement. (1946)

[245] STOUT, Q. F.: Isotonic Regression Algorithms. In: *Retrieved: Aug. 6th* (2011)

[246] SUN, Y. ; KAMEL, M. S. ; WONG, A. K.: Empirical study on weighted voting multiple classifiers. In: *International Conference on Pattern Recognition and Image Analysis* Springer, 2005, S. 335–344

[247] SUTTON, R. S. ; BARTO, A. G. u. a.: *Introduction to reinforcement learning*. Bd. 135. MIT press Cambridge, 1998

[248] SUYKENS, J. A. ; VANDEWALLE, J. : Least squares support vector machine classifiers. In: *Neural processing letters* 9 (1999), Nr. 3, S. 293–300

[249] TABACOF, P. ; COSTABELLO, L. : Probability calibration for knowledge graph embedding models. In: *arXiv preprint arXiv:1912.10000* (2019)

[250] TAN, P.-N. ; STEINBACH, M. ; KUMAR, V. : *Introduction to data mining*. Pearson Education India, 2016

[251] TELEA, A. C.: *Data visualization: principles and practice*. AK Peters/CRC Press, 2007

[252] TENNY, S. ; ABDELGAWAD, I. : Statistical significance. (2017)

[253] THARWAT, A. ; GABER, T. ; IBRAHIM, A. ; HASSANIEN, A. E.: Linear discriminant analysis: A detailed tutorial. In: *AI communications* 30 (2017), Nr. 2, S. 169–190

[254] THORVALDSDÓTTIR, H. ; ROBINSON, J. T. ; MESIROV, J. P.: Integrative Genomics Viewer (IGV): high-performance genomics data visualization and exploration. In: *Briefings in bioinformatics* 14 (2013), Nr. 2, S. 178–192

[255] TRIGUERO, I. ; GARCÍA-GIL, D. ; MAILLO, J. ; LUENGO, J. ; GARCÍA, S. ; HERRERA, F. : Transforming big data into smart data: An insight on the use of the k-nearest neighbors algorithm to obtain quality data. In: *Wiley Interdisciplinary Reviews: Data Mining and Knowledge Discovery* 9 (2019), Nr. 2, S. e1289

[256] TSOUMAKAS, G. ; ANGELIS, L. ; VLAHAVAS, I. : Selective fusion of heterogeneous classifiers. In: *Intelligent Data Analysis* 9 (2005), Nr. 6, S. 511–525

[257] TSUMOTO, S. : Knowledge discovery in medical databases based on rough sets and attribute-oriented generalization. In: *1998 IEEE International Conference on Fuzzy Systems Proceedings. IEEE World Congress on Computational Intelligence (Cat. No. 98CH36228)* Bd. 2 IEEE, 1998, S. 1296–1301

[258] U. REJA, V. H. Katja Lozar Manfreda M. Katja Lozar Manfreda ; VEHOVAR, V. : Open-ended vs. Close-ended Questions in Web Questionnaires. In: *Developments in Applied Statistics* 19 (2003), S. 159–177

[259] VAN ERP, M. ; VUURPIJL, L. ; SCHOMAKER, L. : An overview and comparison of voting methods for pattern recognition. In: *Proceedings Eighth International Workshop on Frontiers in Handwriting Recognition* IEEE, 2002, S. 195–200

[260] VAN SUCH, M. ; LOHR, R. ; BECKMAN, T. ; NAESSENS, J. M.: Extent of diagnostic agreement among medical referrals. In: *Journal of evaluation in clinical practice* 23 (2017), Nr. 4, S. 870–874

[261] VANWINCKELEN, G. ; BLOCKEEL, H. : On estimating model accuracy with repeated cross-validation. In: *BeneLearn 2012: Proceedings of the 21st Belgian-Dutch conference on machine learning*, 2012, S. 39–44

[262] VELLIDO, A. : The importance of interpretability and visualization in machine learning for applications in medicine and health care. In: *Neural Computing and Applications* (2019), S. 1–15

[263] VON DER MALSBURG, C. : The correlation theory of brain function. In: *Models of neural networks*. Springer, 1994, S. 95–119

[264] WAKAP, S. N. ; LAMBERT, D. M. ; OLRY, A. ; RODWELL, C. ; GUEYDAN, C. ; LANNEAU, V. ; MURPHY, D. ; LE CAM, Y. ; RATH, A. : Estimating cumulative point prevalence of rare diseases: analysis of the Orphanet database. In: *European Journal of Human Genetics* 28 (2020), Nr. 2, S. 165–173

[265] WANG, H. ; LEI, Z. ; ZHANG, X. ; ZHOU, B. ; PENG, J. : Machine learning basics. In: *Deep learning* (2016), S. 98–164

[266] WANG, H. ; AZUAJE, F. ; BLACK, N. : An integrative and interactive framework for improving biomedical pattern discovery and visualization. In: *IEEE Transactions on Information Technology in Biomedicine* 8 (2004), Nr. 1, S. 16–27

[267] WARD, M. O. ; GRINSTEIN, G. ; KEIM, D. : *Interactive data visualization: foundations, techniques, and applications*. AK Peters/CRC Press, 2015

[268] WATSON, I. : Case-based reasoning is a methodology not a technology. In: *Research and Development in Expert Systems XV*. Springer, 1999, S. 213–223

[269] WEINREICH, S. S. ; MANGON, R. ; SIKKENS, J. ; TEEUW, M. ; CORNEL, M. : Orphanet: a European database for rare diseases. In: *Nederlands tijdschrift voor geneeskunde* 152 (2008), Nr. 9, S. 518–519

[270] WEST, D. B. u. a.: *Introduction to graph theory*. Bd. 2. Prentice hall Upper Saddle River, 2001

[271] WITTEN, I. H. ; FRANK, E. ; HALL, M. A. ; PAL, C. J.: *Data Mining: Practical machine learning tools and techniques*. Morgan Kaufmann, 2016

[272] WOLD, S. ; ESBENSEN, K. ; GELADI, P. : Principal component analysis. In: *Chemometrics and intelligent laboratory systems* 2 (1987), Nr. 1-3, S. 37–52

[273] WOLPERT, D. H.: Stacked generalization. In: *Neural networks* 5 (1992), Nr. 2, S. 241–259

[274] WOLPERT, D. H.: The lack of a priori distinctions between learning algorithms. In: *Neural computation* 8 (1996), Nr. 7, S. 1341–1390

[275] WOLPERT, D. H. ; MACREADY, W. G.: No free lunch theorems for optimization. In: *IEEE transactions on evolutionary computation* 1 (1997), Nr. 1, S. 67–82

[276] WONG, T.-T. : Performance evaluation of classification algorithms by k-fold and leave-one-out cross validation. In: *Pattern Recognition* 48 (2015), Nr. 9, S. 2839–2846

[277] WOZNIAK, M. ; JACKOWSKI, K. : Some remarks on chosen methods of classifier fusion based on weighted voting. In: *International Conference on Hybrid Artificial Intelligence Systems* Springer, 2009, S. 541–548

[278] XU, H. ; MANNOR, S. : Robustness and generalization. In: *Machine learning* 86 (2012), Nr. 3, S. 391–423

[279] YAN, H. ; DAI, Y. : The comparison of five discriminant methods. In: *2011 International Conference on Management and Service Science* IEEE, 2011, S. 1–4

[280] YE, J. ; JANARDAN, R. ; LI, Q. : Two-dimensional linear discriminant analysis. In: *Advances in neural information processing systems*, 2005, S. 1569–1576

[281] YU, C. ; LIU, J. ; NEMATI, S. ; YIN, G. : Reinforcement learning in healthcare: A survey. In: *ACM Computing Surveys (CSUR)* 55 (2021), Nr. 1, S. 1–36

[282] ZADROZNY, B. ; ELKAN, C. : Obtaining calibrated probability estimates from decision trees and naive bayesian classifiers. In: *Icml* Bd. 1 Citeseer, 2001, S. 609–616

[283] ZADROZNY, B. ; ELKAN, C. : Transforming classifier scores into accurate multiclass probability estimates. In: *Proceedings of the eighth ACM SIGKDD international conference on Knowledge discovery and data mining*, 2002, S. 694–699

[284] ZHANG, P. : Model selection via multifold cross validation. In: *The Annals of Statistics* (1993), S. 299–313

[285] ZHANG, X. ; KLAWONN, F. ; GRIGULL, L. ; LECHNER, W. : VoQs: A Web Application for Visualization of Questionnaire Surveys. In: *International Symposium on Intelligent Data Analysis* Springer, 2015, S. 334–343

[286] ZHANG, Y. ; YANG, Y. : Cross-validation for selecting a model selection procedure. In: *Journal of Econometrics* 187 (2015), Nr. 1, S. 95–112

[287] ZHENG, A. : *Evaluating Machine Learning Models: A Beginner's Guide to Key Concepts and Pitfalls*. O'Reilly Media `https://books.google.de/books?id=OFhauwEACAAJ`. – ISBN 9781491932469

[288] ZHONG, W. ; KWOK, J. T.: Accurate Probability Calibration for Multiple Classifiers. In: *IJCAI*, 2013, S. 1939–1945

[289] ZHOU, Z.-H. : *Ensemble methods: foundations and algorithms*. CRC press, 2012

Curriculum Vitae

	Xiaowei Zhang born on September 19th, 1989 in Gansu, China

Education

04/2015 – 04/2023	**Doctor of Science** in Computer Science Technische Universität Braunschweig, Germany
09/2012 – 06/2014	**Master** in Computer Information System University of Wisconsin, WI, USA
09/2008 – 07/2012	**Bachelor** in Management Information Systems Beijing Information Science and Technology University, China

Professional Experience

10/2015 – 11/2018	**Scientific Researcher** Ostfalia University of Applied Sciences, Germany
09/2014 – 09/2015	**Scientific Researcher** Helmholtz Centre for Infection Research (HZI), Germany
10/2012 – 05/2013	**Student Assistant** in Computer Science Department University of Wisconsin, WI, USA
01/2011 – 09/2011	**Intern** in IT Department COFCO Limited, Beijing, China

List of Scientific Publications

1. Grigull Lorenz, Mehmecke Sandra, Rother Ann-Katrin, Blöß Susanne, Klemann Christian, Schumacher Ulrike, Mücke Urs, **Kortum Xiaowei**, Lechner Werner and Klawonn Frank : **Common pre-diagnostic features in individuals with different rare diseases represent a key for diagnostic support with computerized pattern recognition?** In: PLoS One 14(10), 2019, S. e0222637

2. Sieg Anna-Lena, Das Anibh Martin, Muschol Nicole Maria, Köhn Anja, Lampe Christina, **Kortum Xiaowei**, Mehmecke Sandra, Blöß Susanne, Lechner Werner, Klawonn Frank and Grigull Lorenz: **Künstliche Intelligenz zur diagnostischen Unterstützung ausgewählter seltener lysosomaler Speichererkrankungen: Ergebnisse einer Pilotstudie.** In: Klinische Pädiatrie 231(2), 2019, S. 60 – 66

3. **Kortum Xiaowei**, Grigull Lorenz, Mücke Urs, Lechner Werner and Klawonn Frank : **Improving the decision support in diagnostic systems using classifier probability calibration.** In: International Conference on Intelligent Data Engineering and Automated Learning – IDEAL 2018, S. 419 – 428

4. **Kortum Xiaowei**, Grigull Lorenz, Lechner Werner and Klawonn Frank : **A dynamic adaptive questionnaire for improved disease diagnostics.** In: Advances in Intelligent Data Analysis XVI: 16th International Symposium, IDA 2017, S. 162–172

5. Mücke Urs, Klemann Christian, Baumann Ulrich, Meyer-Bahlburg Almut, **Kortum Xiaowei**, Klawonn Frank, Lechner Werner M and Grigull Lorenz: **Patient's experience in Pediatric Primary Immunodeficiency Disorders: computerized classification of questionnaires.** In: Frontiers in immunology 8(7), 2017, S. 384

6. Blöß Susanne, Klemann Christian, Rother Ann-Katrin, Mehmecke Sandra, Schumacher Ulrike, Mücke Urs, Mücke Martin, Stieber Christiane, Klawonn Frank, **Kortum Xiaowei**, Lechner Werner and Grigull Lorenz : **Diagnostic needs for rare diseases and shared prediagnostic phenomena: Results of a German-wide expert Delphi survey.** In: PLoS One 12(2), 2017, S. e0172532

7. **Kortum Xiaowei**, Klawonn Frank, Grigull Lorenz, Mücke Urs and Lechner Werner : **Diagnosis support for orphan diseases: a case study using a classifier fusion method.** In: International Conference on Intelligent Data Engineering and Automated Learning – IDEAL 2016, S. 379 – 385

8. Grigull Lorenz, Lechner Werner, Petri Susanne, Kollewe Katja, Dengler Reinhard, Mehmecke Sandra, Schumacher Ulrike, Lücke Thomas, Schneider-Gold Christiane, Köhler Cornelia, Güttsches Anne-K, **Kortum Xiaowei** and Klawonn Frank: **Diagnostic support for selected neuromuscular diseases using answer-pattern recognition and data mining techniques: a proof of concept multicenter prospective trial.** In: BMC medical informatics and decision making 16(1), 2016, S. 1 – 9

9. **Zhang Xiaowei**, Klawonn Frank, Grigull Lorenz and Lechner Werner: **VoQs: a web application for visualization of questionnaire surveys.** In: Advances in Intelligent Data Analysis XIV: 14th International Symposium, IDA 2015, S. 334 – 343